T0172943

Kinanthropometry IX

The International Society for the Advancement of Kinanthropometry held its 9th International Conference in Thessaloniki, Greece in August 2004. The meeting was held in conjunction with the 2004 Pre-Olympic Congress, immediately prior to the XXVIII Olympic Games in Athens. This volume contains a selection of papers presented to the Conference, covering the following topic areas:

- three-dimensional whole-body scanning
- body morphology
- bioelectrical impedance analysis
- body composition
- body image

These papers represent the current state of research and knowledge in kinanthropometry, and will be of particular interest to students and researchers in sport and exercise science, kinanthropometry, physical education and human sciences.

Michael Marfell-Jones is Professor and Chair of Health Sciences at the Universal College of Learning, Palmerston North, New Zealand and President of the International Society for the Advancement of Kinanthropometry. **Arthur Stewart** is a Senior Lecturer in the School of Health Sciences at Robert Gordon University, Aberdeen, UK. **Tim Olds** is an Associate Professor in the School of Health Sciences at the University of South Australia.

Kinanthropometry IX

Proceedings of the 9th International
Conference of the International Society for
the Advancement of Kinanthropometry

Edited by

**Michael Marfell-Jones,
Arthur Stewart and Tim Olds**

Routledge
Taylor & Francis Group

LONDON AND NEW YORK

First published 2006 by Routledge
2 Park Square, Milton Park, Abingdon, Oxon OX14 4RN

Simultaneously published in the USA and Canada
by Routledge
270 Madison Ave, New York, NY 10016, USA

Routledge is an imprint of the Taylor & Francis Group, an informa business

© 2006 Michael Marfell-Jones, Arthur Stewart and Tim Olds for editorial
material and selection. Individual chapters © the contributors.

Publisher's Note
This book was prepared from camera-ready copy supplied by the authors.

Printed and bound in Great Britain by
TJ International Ltd, Padstow, Cornwall

All rights reserved. No part of this book may be reprinted or
reproduced or utilised in any form or by any electronic,
mechanical, or other means, now known or hereafter
invented, including photocopying and recording, or in any
information storage or retrieval system, without permission in
writing from the publishers.

Every effort has been made to ensure that the advice and information in
this book is true and accurate at the time of going to press. However,
neither the publisher nor the authors can accept any legal responsibility
or liability for any errors or omissions that may be made. In the case of
drug administration, any medical procedure or the use of technical
equipment mentioned within this book, you are strongly advised to
consult the manufacturer's guidelines.

British Library Cataloguing in Publication Data
A catalogue record for this book is available from the British Library

Library of Congress Cataloging in Publication Data
International Society for the Advancement of Kinanthropometry.
    International Conference (9th : 2004 : Thessalonike, Greece)
    Kinanthropometry IX : proceedings of the 9th International Conference
    of the International Society for the Advancement of Kinanthropometry /
    edited by Michael Marfell-Jones, Arthur Stewart, and Tim Olds.
        p.   cm.
    Includes bibliographical references and index.
    ISBN 0-415-38053-7 (hardback)
    1. Sports—Physiological aspects—Congresses.   2. Kinesiology—
    Congresses.   3. Anthropometry—Congresses.   4. Somatotypes—
    Congresses.   I. Marfell-Jones, Mike.   II. Stewart, Arthur, 1958–.
    III. Olds, Tim.   IV. Title.
    [DNLM: 1. Anthropometry—Congresses.   2. Sports—Congresses.
    3. Body Composition—Congresses.   4. Body Constitution—
    Congresses.   5. Electric Impedance—Congresses.
    QT 260 I59k 2006]
    RC1235.I525 2004
    613.7—dc22
                                                          2005031176

ISBN 0-415-38053-7

Contents

Preface   *vii*

Introduction   ix
M. MARFELL-JONES, A.D. STEWART and T. OLDS

1   **The use of 3D whole-body scanners in anthropometry**   1
     T. OLDS and F. HONEY

2   **Comparative morphology of strongmen and bodybuilders**   15
     A.D. STEWART and P. SWINTON

3   **Built for Success: Homogeneity in Elite Athlete Morphology**   29
     T.R. ACKLAND

4   **A kinanthropometric profile and morphological prediction**   39
     functions of elite international male javelin throwers
     A. KRUGER, J.H. DE RIDDER, H.W. GROBBELAAR and
     C. UNDERHAY

5   **Athletic Morphology: Approaches and limitations using dual X-ray**   53
     absorptiometry and anthropometry
     A.D. STEWART

6   **Monitoring exercise-induced fluid losses by segmental bioelectrical** 65
     impedance analysis
     A. STAHN, E. TERBLANCHE and G. STROBEL

7   **Anthropometric Measurements in Zambian Children**   95
     Z. CORDERO-MACINTYRE, R. DURAN, S. METGALCHI,
     M. RIVERA and G. ORMSBY

8 **Pubertal Maturation, Hormonal Levels and Body Composition** 109
   in Elite Gymnasts
   P. KLENTROU, A.D. FLOURIS and M. PLYLEY

9 **Body Composition Before and After Six Weeks Pre-season** 123
   Training in Professional Football Players
   E. EGAN, T. REILLY, P. CHANTLER and J. LAWLOR

10 **Body image and body composition differences in Japanese** 131
   and Australian males
   M. KAGAWA, D. KERR, S. S. DHALIWAL and C. W. BINNS

11 **The observed and perceived body image of female** 145
   Comrades Marathon athletes
   N. M. BEUKES, R. L. VAN NIEKERK and A. J. J. LOMBARD

Index 155

Preface

The International Society for the Advancement of Kinanthropometry (ISAK) is an international group that has made its role the stewardship and advancement of kinanthropometry - the study of human size, shape, proportion, composition, maturation and gross function. Kinanthropometry, a discipline of only some forty years standing, is named from the Greek root words kinein (to move), anthropos (man) and metrein (to measure). Its purpose is to facilitate a better understanding of growth, exercise, nutrition and performance. Although ISAK is probably best known for its international anthropometry accreditation scheme, it also plays a major role in increasing international dialogue between kinanthropometrists and facilitating the presentation and dissemination of the results of their scientific endeavours.

To that end, for the past eighteen years, ISAK has made a practice of holding its Biennial General Meetings in association with major sport science conferences prior to either the Olympic or Commonwealth Games. Within these conferences, ISAK has been instrumental in the organisation of a kinanthropometry stream as a platform for the presentation of quality papers on the research findings of its members and international colleagues. In the past decade, that has occurred in Dallas in 1996 (prior to the Atlanta Olympics); in Adelaide in 1998 (prior to the XVI Commonwealth Games in Kuala Lumpur); in Brisbane in 2000 (prior to the Sydney Olympics); and in Manchester (prior to the XVII Commonwealth Games held there). Subsequent to these conferences, ISAK appointed editors to produce a set of proceedings so that at least some of the papers could be accessed by a far wider audience than just those able to attend the conferences.

In August 2004, this tradition continued in Thessaloniki, Greece, in conjunction with the 2004 Pre-Olympic Congress, immediately prior to the Athens Olympics. These proceedings, which represent a refereed selection of the best kinanthropometry papers presented in Thessaloniki, are the ninth in the series. ISAK thanks the editors most sincerely for their efforts in compiling this volume and congratulates the authors on their work and its inclusion in this publication.

Mike Marfell-Jones
President
International Society for the Advancement of Kinanthropometry

Introduction

Michael Marfell-Jones[1], Arthur Stewart[2] and Tim Olds[3]
[1]UCOL, Private Bag 11 022, Palmerston North, New Zealand
[2]School of Health Sciences, The Robert Gordon University, Garthdee
Road, Aberdeen, AB107QG, United Kingdom
[3]Centre for Applied Anthropometry, University of South Australia,
City East Campus, GPO Box 2471, Adelaide SA 5001, Australia

The contents of this volume represent the Proceedings of Kinanthropometry IX, the 9th International Conference in the subject area of kinanthropometry. The conference was incorporated within the 2004 Pre-Olympic Congress in Thessaloniki, Greece from 6 to 11 August 2004, the week preceding the Athens Olympic Games. The kinanthropometry strand of the conference was organised by the International Society for the Advancement of Kinanthropometry (ISAK) in conjunction with the conference hosts, the Department of Physical Education and Sport Science of the Aristotle University of Thessaloniki, under the guidance of the Organising Committee Secretary-General, Spiros Kellis.

Following the Conference, all presenters whose papers had a kinanthropometric focus were invited to submit full manuscripts for consideration for selection. Not all communications were worked up into manuscripts for the Proceedings and not all of those that were written up succeeded in satisfying peer reviewers.

The selected chapters provide a good selection from the kinanthropometric spectrum. The opening chapter, on the use of 3D whole-body scanners to measure size and shape, is followed by four papers on body morphology, two papers on bioelectrical impedance analysis, two papers on body composition and, finally, two papers on body image. The international contributions both provide readers with an outline of the current state of knowledge in kinanthropometry around the globe and reflect a broad interest in the subject area world wide. Therefore, the book should have a large appeal across a number of countries.

The editors are grateful to contributors for their positive involvement in the compilation of this book, particularly in responding to the comments from the review process. We also acknowledge the invaluable technical assistance of Ms Jennifer Gordon, Faculty of Health Sciences at the Universal College of Learning, for compiling the camera-ready copy of the text.

Michael Marfell-Jones
Arthur Stewart
Tim Olds

CHAPTER ONE

The use of 3D whole-body scanners in anthropometry

Tim Olds and Fleur Honey
Centre for Applied Anthropometry
University of South Australia

1. INTRODUCTION

Modern surface anthropometry evolved from late nineteenth-century anthropology. It was heavily bone-based, static and taxonomic. Many of our current anthropometric practices – the logic of landmarking, for example – descend in a direct line from this tradition. This tradition is not well adapted to modern anthropometric disciplines — ergonomics and apparel design, computer simulations and biomedical research — which require estimates of soft-tissue dimensions, shape characteristics, functional and dynamic properties, and which can handle and analyse gigantic datasets.

In the past, our limitations have been partly ideological, and partly technological. Restricted to the linearity of the anthropometer and the tape, we have been tied to one-dimensional constructions of the body. Lacking methods of three-dimensional capture and analysis, we have largely neglected shape elements and tissue topography. We have lacked the morphometric tools to describe body plasticity associated with maturation, ageing, training and evolution. Above all, we have lacked the communications systems to ensure standardisation and to build accessible data repositories.

Today, however, we have those tools: 3D whole-body scanners, axial tomography and magnetic resonance imaging, motion capture systems, sophisticated software for morphometric analysis and internet-based data warehouses. Unfortunately, anthropometric vision has not kept pace with anthropometric tools. This paper describes one of these new tools — 3D whole-body scanners — and suggests some ways in which they can be applied in a new anthropometry.

Whole-body scanners are relatively new. Although there were commercial and research models available in the 1980s, it was only in the latter half of the 1990s that hardware-software suites became accurate, reliable and cheap enough to be widely used for the collection of anthropometric data (Istook and Hwang, 2001). Manufacturers of 3D whole-body scanners claim that they remove some of the greatest problems associated with anthropometric surveys, especially time and data reproducibility (Mckinnon and Istook, 2001). Whole-body scanners make possible a much larger measurement set, including surface and cross-sectional areas and volumes.

Section 2 of this paper will describe how 3D whole-body scanners work. Section 3 looks at current and potential applications of 3D scan data. Section 4 reports on studies of the accuracy and reliability of 3D scanners. Section 5 concludes with a look towards the future.

2. HOW SCANNERS WORK

2.1 Scanning hardware

Three-dimensional (3D) scanners capture the outer surface of the body using a light source and light-sensitive devices. Light is shone onto the body, and the pattern it makes with the body is captured by multiple cameras. This pattern is then decoded by a computer to recreate a virtual model of the three-dimensional body surface.

The first scanners used a shadowing method, where a light source was shone onto a body, generating a shadow projected onto a chequerboard. This was captured by a camera. The body was then rotated slightly, and the procedure repeated. The captured silhouettes were assembled to create a 3D model of the body surface (Istook and Hwang, 2001).

In one of the earliest commercial scanners, the Loughborough Anthropometric Shadow Scanner (LASS), a vertical slit of light was projected onto the scanned body and an off-axis camera detected the deformation of the slit, caused by the shape of the body. Three-dimensional co-ordinates were calculated for each point of the light slit on the body using triangulation. In order to capture the whole body, the subject stood on a rotating platform, which was turned 360° in measured increments (Buxton, Dekker, Douros and Vassilev, 2000).

Current 3D whole-body scanning systems also use a light source (laser or white light) to project straight lines or grids onto the scanned body. Most use a horizontal beam which moves rapidly down the body in about 10 s. Different angles are captured by charge coupled device (CCD, ie digital) cameras placed all around the body, so that the subject does not need to be rotated. Although the beam appears to pass smoothly and continuously down the body, it is actually projected repeatedly and very rapidly at intervals of about 3-5 mm, so that the body appears as a series of points on evenly-spaced lines. The distance between the lines is called the "pitch" of the device. CCD cameras detect the deformed light stripe against a grid reference. The Z co-ordinate (height) of all points along that stripe are identical. The Y (antero-posterior) co-ordinate for any point on the line is determined by triangulation as a triangle is formed by the light source, body and camera. The distance between the light source and camera is known, and the angle between the camera and the reflected light from the body can be measured. As a result, the distance from the laser to the body can be calculated, giving the Y co-ordinate. The X (side-to-side) co-ordinate is calculated relative to the reference grid.

White light scanners are generally cheaper and faster, but tend to produce lower quality scans, sometimes with large gaps. These systems project a grid of light onto the body rather than a single line. Once again, the deformation of this grid as reflected off the 3D body surface is captured by CCDs. The deformed grid is compared to an orthogonal reference grid, and the interaction of the two grids causes interference or "moiré-fringe" patterns. Some white-light scanners move the light grid by preset increments to improve capture resolution, a process known as "phase measuring profilometry" (PMP).

Scanners are quite large — roughly the size of a telephone booth. They can be moved, but take several hours to assemble and calibrate. Typical costs range from about $US50,000 to $US250,000, including software. Table 1 shows some of the major manufacturers currently producing 3D whole-body scanners.

Scanning System	Product	Light Source
Cyberware	WBX, WB4	Laser
Hamamatsu	Bodyline scanner	Laser
Hamano	Voxelan	Laser
TC^2 Image Twin	2T4	Light
Vitronic	Vitus Smart, Vitus Pro	Laser

Table 1. Some of the major 3D scanning systems in use today (Simmons and Istook, 2003).

2.2 Scanning software

2.2.1 Primary processing

Subjects usually wear form-fitting briefs during the scanning process and adopt a standardised pose (Istook and Hwang, 2001). Scanners reproduce the body surface as a "point cloud", typically consisting of about 600,000 points, each represented by an XYZ co-ordinate. Point cloud data require quite intensive processing before they are usable for anthropometry. Primary filtering eliminates outliers caused by reflected light. Because scanners use a number of cameras (typically eight) to capture different views of the body, the data from each camera must be combined to produce a complete three-dimensional view of the scanned body. This process of "stitching" together to create a full 3D image is called "registration" (Paquette, 1996). These data manipulations are generally performed by software which is native to the scanner, such as Human Solutions' *ScanWorX* (native to Vitus scanners) or Hamamatsu's *BodyLine Manager*.

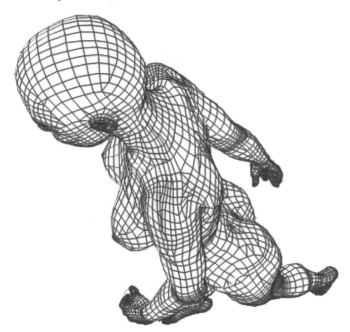

Figure 1. A polygonal mesh or wireframe figure.

To fully exploit the measurement capabilities of 3D scanners, neighbouring points must be joined to form a polygonal mesh or "wireframe" (Figure 1). This facilitates the calculation of areas and volumes. The body surface captured by a scanner always has gaps and missing areas caused either by parts of the body shadowing others parts (eg under the arms), light striking the body at almost 180° (top of the head), or reflection off complex surfaces (hair). Some scanning software, such as the Hamamatsu's *BodyLine Manager* or Cyberware's *CySize*, will "autofill" these gaps intelligently, producing "water-tight" surfaces. Wireframes can be "rendered" to produce smooth surfaces (the metaphor is from plastering). Figure 2 shows an example of a scanned body rendered in the figure-posing software *Poser*.

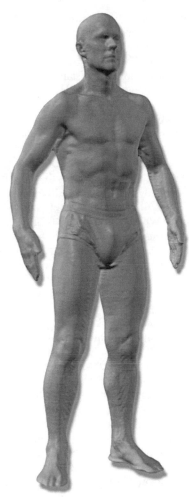

Figure 2. A scanned body "rendered" in the figure-posing software *Poser*.

There is a confusing array of different 3D file formats. For whole-body scanning, two of the most common formats are .PLY and .OBJ. Commercial and freeware translators, such as *MTool*, make it possible to translate into and from many of these formats. Frustratingly, however, some manufacturers still use proprietary file formats which can only be translated by their own software.

2.2.2 Measurement extraction

In order to extract measurements from the scan data, special measurement extraction software is required. Some of the more commonly used systems are Cyberware's *DigiSize* and *CySize*, Hamamatsu's *BodyLine Manager*, Human

Solutions' *ScanWorX*, and *NatickMsr*, developed by the US Soldier Center at Natick, MA. These programs can extract a very large number of traditional measurements (such as biacromial breadth and waist girth), in relation to a landmarking system (Istook and Hwang, 2001). Measurement extraction software also allows the measurement of contour distances, cross-sectional areas, surface areas and volumes, thereby greatly expanding the range of applications 3D scanning has to offer (Paquette, 1996).

For standardisation, all measurements must refer to a landmark system. Unfortunately, different software suites tend to use slightly different or ill-defined landmarks. There are many different competing and overlapping landmark systems, including those of the International Society for the Advancement of Kinanthropometry (ISAK, 2001), the Civilian And European Surface Anthropometric Resource (CAESAR; Blackwell, et al., 2002), the US Army's Anthropometric Survey (ANSUR; Clauser, et al., 1988), the International Biological Program (IBP; Weiner and Lourie, 1969), and the International Standards Organization clothing specifications (ISO, 1996).

Whatever system is used, measurement extraction software must be able to locate landmarks. The underlying bony points used as landmarks are not scanned, due to 3D scanners only generating a surface image. Landmarks can be identified in three ways (Buxton, Dekker, Douros, and Vassilev, 2000):

1) *Automatic landmark recognition* (ALR) uses software to identify landmarks from the scan data, without human intervention. Two methods for automatic landmark identification are template matching, and curvature calculation (Suikerbuik, Tangelder, and Daanen, 2004). Template matching requires creation of a base template, with all points of interest marked. This template is then deformed to fit the scan data. Once deformed, the XYZ co-ordinates of the points of interest are recorded. Pattern recognition, or curvature calculation, detects features based on surface shape. An area of the scan is localized and then a search is performed to find the point based on slopes and gradients (Buxton et al., 2000). In its current state of development, ALR has been shown to be unacceptably inaccurate for many purposes. It assumes constancy of bony alignment throughout the population and does not acknowledge variation in shapes and sizes (Suikerbuik et al., 2004). ALR software also often allows only pre-set measurements.

2) *Digital Landmark Recognition* (DLR) involves landmarks being located by an operator on the scanned image. DLR can be both accurate and reliable on thin subjects. However, it is often difficult to locate bony landmarks on fat or very muscular subjects as underlying points are covered with subcutaneous fat, or muscle tissue (Buxton et al., 2000). DLR is suitable for some easily-recognisable landmarks such as the Dactylion or Malleoli.

3) *Physical Landmark Recognition* (PLR) is where landmarkers (a "landmarker" refers to a physical marker placed at a landmark site) are physically placed on the subject's body prior to scanning. Some scanning systems, with colour or texture recognition capabilities, can automatically recognise special landmarkers. With others, landmarkers need to be raised and identified on the scanned image.

Due to the inaccuracy of ALR and the difficulty with DLR on fat or muscular subjects, PLR is preferred as the most accurate, though most time-consuming, method (Suikerbuik et al., 2004).

Whatever landmark recognition system is used, identified landmarks are coded as XYZ co-ordinates and used as anchor points in measurement determination. Linear distances are calculated as point-to-point measures. Contour measurements (along the surface of the body) are calculated as the intersection between the body surface and a plane passing through specified landmarks at a certain orientation. Surface areas can be calculated by summing the areas of polygons on a polygonal mesh, and volumes can be calculated either by integration of the cross-sectional area of slices, or by summing the volumes of triangular prisms. Some applications have the ability to simulate "tape-measure" dimensions which do not follow the skin exactly (Figure 3).

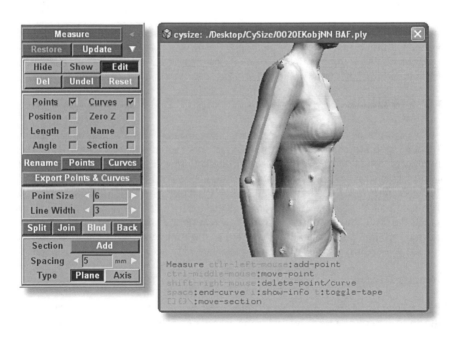

Figure 3. An example of measurement extraction software. *CySize* has the ability to calculate point-to-point measurements, contours and tape measure girths.

3. APPLICATIONS OF 3D SCANNING SYSTEMS

Three-dimensional body scanning methods are currently being used in areas such as health, body image research, human factors, graphic design, and apparel design and manufacture (Mckinnon and Istook, 2001).

3.1 Health

In the health field, 3D scanning has been used for cranio-facial and reconstructive surgery, as well as the manufacture of prostheses, where the undamaged limb is scanned and a mirror image is produced for the damaged side. Along with this, scanning has been used to calculate wound volume and burn surface areas, to assist with planning therapy and assess healing. Scan-derived volumes can be used to assess œdema, changes with the menstrual cycle, and changes in pregnancy. There is potential for 3D scanning to be used to find better anthropometric predictors of disease status or risk. At the moment, fairly crude manual measurements are made to find waist and hip girth cut-offs for diabetes risk. Three-dimensional scanning allows a wide range of abdominal cross-sectional areas and volumes to be calculated as well.

3.2 Graphic design

Graphic designers are using 3D body scanners to create "avatars", computer-generated human-like figures that can feature in video games, movies and simulations. Three-dimensional scans of the Governor of California have been used to create his character in the *Terminator* video-games.

3.3 Human factors

Software programs called "human modeling programs" (HMPs) can use data from 3D scans to rescale human-like "manikins" or virtual humans, which are then animated to interact with a digital environment, to aid in design of workspaces, transportation and seating. Some of the more common HMPs are *Jack*, *Ramsis*, *Safeworks* and *SAMMIE*. We are currently working with the Australian Defence Force to develop anthropometric profiles required for pilots to fly in a range of aircraft. Aircrew and potential recruits are scanned, as are crewstations. Both are imported into *Jack*, and manikins rescaled to the dimensions of individual aircrew are made to perform a series of critical tasks to verify accommodation and reach.

3.4 Sports science

3D body scans of swimmers have been used by the Western Australian Sports Institute to calculate hydrodynamic characteristics. The same methodology could be applied in calculating the projected frontal area of cyclists, a major determinant of air resistance. Currently this is done by a time-consuming manual method. Calculation of limb and truncal volumes could be used to monitor the effectiveness of dietary and training interventions. There have been some studies of the possibility of using 3D scanning to estimate body volume, and hence body density and body composition, but the level of accuracy is not yet sufficient.

3.5 Apparel

The apparel industry has been the main driving force behind the development and improvement of 3D body scanning because of the range of potential applications that it has. The simplest application is inventory control. An extensive list of measurements can be taken from a 3D body scan, such as sleeve in-seam and waist girth. These measurements can then be compared with existing clothing size templates to determine the percentage of people wearing each size, thereby allowing manufacturers and retailers to choose appropriate inventories for their target population (Istook and Hwang, 2001).

An extension of this is shape analysis. Measurements extracted from the scans can be used to identify "clusters" of body shapes. Shape categorisation may use statistical techniques such as principal components analysis or cluster analysis. This would allow the development of new size templates, thus increasingly the likelihood of a good fit (Kaufmann, 1997).

The end goal of research into 3D body scanners and the apparel industry is "mass customisation" of garments, whereby "perfect fit" garments are created that mould to the unique 3D shape of each individual (Carrere, Istook, Little, Hong and Plumlee, 2002). Mass customisation would lead to better fitting garments, decrease the amount of work for clothing manufacturers, decrease inventory and decrease the number of returns (due to poor fit), hence reducing costs (Paquette, 1996; Kaufmann, 1997). "Virtual try-on" technology involves images of garments being superimposed on your personal 3D image, much like dressing a 3D doll of yourself. It allows the customer to see what a garment looks like when worn, without actually trying it on (Cordier et al., 2003). The European *eTailor* project is exploring links between 3D scanning and clothing manufacture. It aims to find a solution to the fitting problem, to enable the production of "perfect fit" garments, and to develop virtual shopping.

4. ACCURACY AND PRECISION

The viability of these applications depends on the accuracy and reliability of measurements derived from 3D scans. There have been very few published studies of measurement error in comparing physical measurements to measurements derived from scans. They are not always comparable, using different hardware-software suites and reporting measurement error using different statistics. Here we will summarise those studies, attempting to translate their results into a common statistical metric. Bland-Altman analysis quantifies both systematic error or bias (defined as the mean of the differences between physical and scan-derived measurements), and random error, quantified here as the Standard Error of Measurement (SEM), which is the standard deviation of the differences divided by $\sqrt{2}$. When the bias is zero, the SEM is the same as the Technical Error of Measurement.

4.1 Accuracy

An in-house study from the manufacturers of TC^2 scanners reported biases of –47.0 to –27.9 mm (scanned dimensions were greater than physical dimensions). No SEMs were reported. Brooke-Wavell, Jones and West (1994) compared physical measurements to measurements derived from physically landmarked scans from the LASS whole-body scanner. SEMs ranged from 1.3 to 10.6 mm, with a mean of 5.2 mm. There were, however, substantial biases, ranging from –27.8 to +16.8 mm. Mckinnon and Istook (2002) compared physical measurements to measurements derived from scans using Automatic Landmark Recognition (ALR; TC^2 system). The SEMs ranged from 3.2 to 11.5 mm, while biases averaged ±19.3 mm. Some of this error was presumably due to the ALR algorithms not identifying landmarks corresponding to the measurer-located landmarks. Our recent study revealed SEMs of 2.2 to 9.8 mm for the Hamamatsu scanner, and biases of –25.1 to +12.8 mm, and similar values for the Vitus Smart. Both studies used physical landmark recognition. The average absolute bias for the Vitus Smart was 5.5 mm, and the mean SEM was 5 mm, or about 1-2% of mean dimensions. These results are summarised in Table 2.

Reference	Scanner	Bias (mm)	SEM (mm)
McKinnon and Istook (2002)	TC^2 (ALR)	+/- 19.3[a]	3.2 to 11.5
Brooke et al. (1994)	LASS	–27.8 to +16.8	1.3 to 10.6
TC^2 (in-house)	TC^2 (ALR)	–47.0 to –27.9	NA
Present study	Hamamatsu	–25.1 to +12.8	2.2 to 9.8
Present study	Vitus Smart	–28.0 to +15.2	2.1 to 9.4

Table 2. Summary of studies of the accuracy of 3D whole-body scanners, relative to physical measurement.

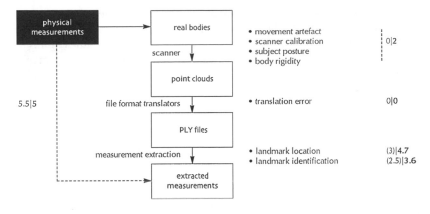

Figure 4. Assumptions and possible sources of error in moving from real world bodies to dimensions extracted from scans. The measured or estimated errors (biases|SEMs) in mm are also shown. Derived values are in parentheses.

Our study also identified the errors associated with different steps in the process of scanning and measurement extraction. These are summarised in Figure 4. The random error in moving from real bodies to point clouds is probably <2 mm (this is mainly due to scanner calibration), and there is no systematic bias. There appears to be no error associated with file format translation. Most error is introduced in landmark location on the physical body, and the identification of landmarkers on the scanned image. Intra- and inter-tester error in landmark location on the body was assessed by ISAK-accredited anthropometrists using an ultraviolet pen, so that previous locations would not bias subsequent ones. The mean intra-tester SEM across 22 sites was 3.5 mm, and the mean inter-tester SEM was 6.0 mm.

Intra- and inter-tester error in identifying landmarkers on the scanned image was assessed by having six anthropometrists identify twice the XYZ co-ordinates of 43 landmarks on the scans of 10 subjects. For physically landmarked sites, the intra-tester SEM was 2.2 mm, and the inter-tester TEM 3.8 mm. For digitally located landmarks, the mean intra-tester SEM was 2.6 mm, and the inter-tester TEM 5.3 mm.

4.2 Precision

When the same scans were analysed twice by the same anthropometrist and measurements were extracted, the mean bias was 0.4 mm, and the mean SEM was 2.7 mm (1.4% of mean values). When the same scans were analysed by two different anthropometrists and measurements extracted, the mean bias was 0 mm, and the mean SEM 3.1 mm (1.5% of mean values).

These studies suggest that when used by experienced and trained anthropometrists, 3D scanning can yield linear dimensions which are both precise, and accurate relative to physical measurements. There have been fewer still studies of the accuracy and precision of area and volume determinations using 3D scanning.

5. CONCLUSION

While 3D whole body scanning has a number of limitations relative to traditional surface anthropometry — expense, technical expertise required, inability to take skinfolds and compressed bone breadths — it offers an opportunity to collect much greater volumes of data without the subject being present during measurement. It also expands the types of measurements which can be made into two and three dimensions. It captures anthropometric data in a form which can be fed into and used by new and powerful software applications in areas such as ergonomics and apparel design.

Acknowledgements

This paper draws on work supported by the Commonwealth of Australia, RAAF Project MIS872. The assistance of the Australian Defence Force is gratefully acknowledged. The authors acknowledge the contribution of Sincliar Knight Merz, the Australian Institute of Sport, and the University of Ballarat to the project.

REFERENCES

Blackwell, S., Robinette, K.M., Boehmer, M., Fleming, S., Kelly, S., Brill, T., Hoeferlin, D., Burnsides, D., and Daanen, H., 2002, *Civilian American and European Surface Anthropometry Resource (CAESAR), Final Report, Volume II: Descriptions*. Warrendale, PA: United States Air Force Research Laboratory.

Brooke-Wavell, K., Jones, P.R.M., and West, G.M., 1994, Reliability and repeatability of 3-D body scanner (LASS) measurements compared to anthropometry. *Annals of Human Biology, 21*, pp.571-577.

Buxton, B., Dekker, L., Douros, I., and Vassilev, T., 2000, *Reconstruction and Interpretation of 3D Whole Body Surface Images.* (London: University College London).

Carrere, C., Istook, C., Little, T., Hong, H., and Plumlee, T., 2002, *Automated garment development from body scan data*. North Carolina State University: National Textile Centre Annual Report.

Clauser, C., Tebbets, I., Bradtmiller, B., McConville, J., and Gordon, C.C., 1988, *Measurer's handbook: U.S. Army anthropometric survey 1987-1988*. Natick, MA: United States Army Natick Research, Development and Engineering Center.

Cordier, F., Seo, H., and Magnenat-Thalmann, N., 2003, Made-to-measure technologies for an online clothing store. *IEEE Computer Graphics and Applications, 23*, pp.38-49.

International Society for the Advancement of Kinanthropometry, 2001, *International standards for anthropometric assessment*. Adelaide, SA: Author.

International Standards Organization, 1996, *Basic human body measurements for technological design. TC 159/SC3*. Geneva: Author.

Istook, C., and Hwang, S., 2001, 3D body scanning systems with application to the apparel industry. *Journal of Fashion Marketing and Management, 5*, pp.120-132.

Kaufmann, K., 1997, Invasion of the body scanners. *Circuits and Devices, 13*, pp.12-17.

Mckinnon, L., and Istook, C., 2001, Body scanning. Comparative analysis of the ImageTwin system and the 3T6 Body Scanner. *Journal of Fashion Marketing and Management, 6*, pp.103-121.

Mckinnon, L., and Istook, C., 2002, Body scanning. The effects of subject respiration and foot positioning on the data integrity of scanned measurements. *Journal of Textile and Apparel Technology and Management, 1*. pp.

Paquette, S., 1996, 3D scanning in apparel design and human engineering. *IEEE Computer Graphics and Applications, 16*, pp.11-15.

Simmons, K., and Istook, C., 2003, Body measurement techniques, comparing 3D body-scanning and anthropometric methods for apparel applications. *Journal of Fashion Marketing and Management, 7*, pp.306-332.

Suikerbuik, R., Tangelder, H., and Daanen, H., 2004, Automatic feature detection in 3D human body scans. *Proceedings of the SAE Digital Human Modelling for Design and Engineering Conference, June 15-17, Rochester, England*. Rochester: SAE.

Weiner, J.S., and Lourie, J.A., 1969, *Human biology: a guide to field methods (IBP handbook no. 9)*. (Oxford: Blackwell).

CHAPTER TWO

Comparative morphology of strongmen and bodybuilders

Arthur D. Stewart[1] and Paul Swinton[2]

[1]School of Health Sciences, The Robert Gordon University, Aberdeen, UK.

[2]School of Medical Sciences, University of Aberdeen, UK.

1. INTRODUCTION

Since ancient civilisations, humankind has been fascinated by the pursuit of strength. The earliest reference to formal strength training occurs in Chinese texts dating as far back as 3,600 BC when emperors made their subjects exercise daily. There is also evidence that strength training has been performed in several ancient civilisations – perhaps most famously typified by the Greek Milo of Croton, six-times wrestling champion at that the ancient Olympiad who trained by lifting a growing calf which provided an overload effect as it gained weight. Whether for the purposes of warfare, competition or art, amply recorded by the artefacts of the ancient Greek and Roman empires, strength was perceived as highly functional.

Strength training confers considerable morphological adaptation on the body, which can be crafted for aesthetic or functional purposes. Whether the competition rewards the morphology itself, or a physical manifestation of its capability, strength sports have existed in more recent history since the late 19th century. Demonstrations of absolute strength were made public by early showmen in the travelling circus. These included such feats as those of Eugene Sandow (1867-1925) billed as 'the world's most perfectly developed man' who organised bodybuilding competitions in various regions of Britain, culminating in a final at London's Royal Albert Hall in 1901 attended by 15000 people. The history of the "strongman" discipline is less clear, drawing from various historical and other contemporary influences in several countries, such as Scottish Highland games and Basque contests in Spain. Specific tasks included conventional weight lifting, or pushing, pulling, carrying or throwing heavy objects, in which the time or distance was scored. In recent decades both bodybuilding and strongmen sports have become established in highly competitive circuits and organisational structures world-wide.

Bodybuilders have different approaches to training from strongmen, because the strength they acquire is not assessed in the performance, which is based on aesthetic factors associated with the physique itself. For competition, the athletes minimise subcutaneous adiposity to render the underlying muscle 'topography' more apparent to the observer.   Such interventions, usually achieved by a combination of extreme diet and exercise behaviours, are associated with substantial fat and some lean tissue losses (Goris and Somers, 1986; Huygens et al., 2002).   In contrast strongmen do not compete in weight categories, and require a great absolute strength, with little need to be concerned over appearance, total mass or fat mass.   To date strongmen have never had anthropometric profiles reported.   Therefore it was the purpose of this study firstly to report the morphology of strongmen in a comprehensive and standardised way, and secondly to explore similarities and differences in physique between strongmen and bodybuilders.

2. METHODS

2.1 Sample

Twenty apparently healthy male athletes who competed either in bodybuilding (n = 10) or strongman (n = 10) competition were included in the study.   The strongmen and bodybuilders competed either at regional, national and international level.   No athletes were scheduled to compete for the following two months and it was anticipated that they would be at their heaviest at the time of measurement. They were measured by anthropometrists accredited by the International Society for the Advancement of Kinanthropometry (ISAK), at the University of Aberdeen, UK, overseen by a Level 4 ISAK criterion anthropometrist.   All measurements were made subject to informed consent and local ethical approval.   The study was conducted in University laboratories and local gymnasia.   A full protocol of 39 measurements was made on each subject, with two measurements for each variable measured if repeated measurement targets (5% for skinfolds and 1% for all other measures) were achieved, and three measurements, if not.   Calculations were based on the mean of two, or the median of three measures for each variable in accordance with ISAK procedures (ISAK, 2001).

2.2 Measurements

Measurements were made when subjects had avoided eating for 2-3 hours, but were fully hydrated and recovered from previous exercise.   Total body mass was measured to 100 g on a Seca Omega 873 digital floor scale (Hamburg, Germany) and stretch stature was obtained using a Holtain stadiometer (Crymych, UK). Skinfolds were measured by Harpenden Calipers (British Indicators, Luton, UK) at triceps, subscapular, biceps, iliac crest, supraspinale, abdominal, thigh and medial calf sites.   Girths were measured at the calf (maximum) and arm (flexed and tensed) using a modified Lufkin anthropometric tape (Rosscraft, Surrey, Canada). Breadths were measured at the distal humerus and femur using a Tommy 2 bone caliper (Rosscraft, Surrey, Canada), and at the shoulders, thorax and pelvis using a Campbell 20 wide-spreading caliper (Rosscraft, Surrey, Canada).

Measures were made following ISAK protocols (ISAK, 2001).   Data were recorded onto standard proformas, and entered into SPSS version 9.0 for subsequent analysis.   Anthropometric somatotype was calculated by computer according to established formulae (Duquet and Carter, 2001).   Phantom Z scores were calculated according to the method of Ross and Wilson (1974).   Corrected girths were calculated at thigh, calf and arm by subtracting the skinfold multiplied by *pi* from the girth (Martin et al., 1990)

3. RESULTS

The physical characteristics of both groups are described in Table 1.

Table 1. Physical Characteristics of Subjects

	Bodybuilders	Strongmen
Age (yr)	30.8 (11.5)	29.2 (8.7)
Stature (cm)	176.7 (5.9)	188.2 (4.1)**
Mass (kg)	99.9 (4.4)	126.3 (8.2)**
Body Mass Index (kg.m^{-2})	32.1 (2.4)	35.8 (2.0)*
Σ8 Skinfolds	122.8 (36.0)	174.8 (26.3)*

Independent T tests * P < 0.01 ** P < 0.001

Somatotypes of both groups are depicted in Figure 1 (overleaf).

There was no age difference between groups, but the strongmen group had greater stature and were heavier than bodybuilders (P < 0.001), and had a significantly greater sum of skinfolds (P < 0.01). While both groups would be categorised as endomorphic mesomorphs, somatotype rating differed between groups, primarily due to strongmen being more endomorphic, and to a lesser extent, less ectomorphic (P < 0.001). Both bodybuilders and strongmen had acquired substantial extra muscle, and strong men, some additional fat.

Full anthropometric profiles for each athletic group are reported in Tables 2–5.

Table 2. Skinfolds of strength athletes

	Bodybuilders	Strongmen
Triceps	10.9 (3.7)	16.9 (3.8)
Subscapular	19.3 (3.7)	28.1 (4.4)
Biceps	5.0 (1.5)	8.3 (2.6)
Iliac Crest	24.6 (6.7)	30.4 (3.8)
Supraspinale	12.2 (4.4)	22.3 (5.2)
Abdominal	23.6 (6.5)	32.0 (5.6)
Mid thigh	17.2 (6.7)	22.4 (6.9)
Medial calf	10.6 (4.0)	14.4 (1.9)

All measurements in mm. Mean (SD)

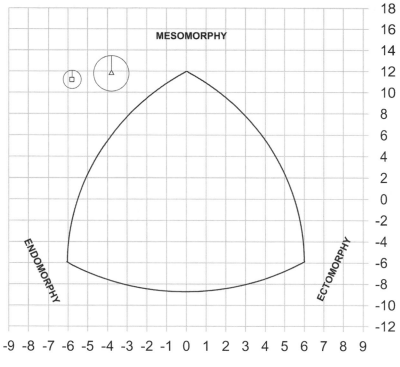

☐ Strongmen   5.9 - 8.6 - 0.1 (0.86)
△ Bodybuilders   4.1 - 8.1 - 0.3 (1.56)

Radii and circles represent somatotype additudinal mean

Figure 1. Somatotypes of bodybuilders and strongmen

Both groups are endomorphic mesomorphs, with the bodybuilders displaying greater physique variation.

Table 3. Girth measurements of strength athletes

	Bodybuilders	Strongmen
Head	56.9 (1.3)	58.5 (1.3)
Neck	42.2 (2.4)	48.0 (1.8)
Relaxed arm	40.1 (3.3)	45.8 (2.7)
Flexed arm	43.7 (3.3)	48.8 (2.2)
Forearm	34.0 (2.3)	37.2 (1.5)
Wrist	18.3 (0.6)	19.1 (1.1)
Chest	115.0 (7.9)	132.1 (6.0)
Waist	95.4 (7.0)	105.7 (3.9)
Gluteal	105.2 (5.4)	116.1 (3.6)
Thigh (1cm)	68.7 (1.6)	74.3 (2.9)
Mid thigh	62.7 (3.8)	63.2 (2.3)
Calf	41.6 (1.2)	45.1 (1.7)
Ankle	22.5 (0.7)	24.6 (1.2)

All measurements in cm. Mean (SD)

Table 4. Segmental lengths of strength athletes

	Bodybuilders	Strongmen
Acromiale-radiale	33.0 (1.7)	35.6 (1.4)
Radiale-stylion	26.1 (0.8)	27.7 (0.8)
Stylion-dactylion	19.5 (0.6)	21.3 (1.1)
Iliospinale height	96.8 (4.0)	103.8 (2.8)
Trochanterion height	91.7 (2.7)	98.5 (2.7)
Troch-tibilale laterale	46.2 (1.4)	48.6 (1.5)
Tibiale laterale ht	45.2 (2.9)	49.6 (1.7)
Tib mediale-sphyrion tibiale	38.7 (1.5)	41.8 (1.3)

All measurements in cm. Mean (SD)

Table 5. Bone breadths of strength athletes

	Bodybuilders	Strongmen
Biacromial	43.5 (1.0)	44.5 (0.7)
Bicristal	29.7 (0.8)	32.9 (2.3)
Foot length	27.1 (1.2)	28.6 (0.9)
Sitting height	95.1 (3.0)	98.6 (2.7)
Transverse chest	32.3 (1.2)	34.6 (1.8)
A-P chest depth	23.2 (1.4)	23.2 (2.4)
Humerus breadth	7.3 (0.3)	7.5 (0.2)
Femur breadth	9.9 (0.2)	10.6 (0.2)

All measurements in cm. Mean (SD)

Universally, the measurements of the strongmen are greater than those of bodybuilders (with the exception of A-P chest depth, which is equal). Establishing whether or not strongmen had greater relative adiposity or muscularity using skinfolds, girths and corrected girths requires a size adjustment whereby both groups are seen as if they were of the same stature. Consequently the profiles of both groups of athletes were calculated using phantom z-scores, for skinfolds, girths, segmental lengths and bone breadths. Phantom z-scores enable a height-independent comparison which accounts for the size difference and are illustrated in figures 2–5 respectively.

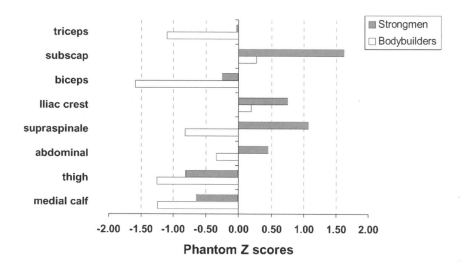

Figure 2. Skinfold Profile

At all sites, strongmen displayed greater values than bodybuilders. The difference between the groups was greatest at the supraspinale, and least at the thigh. These sites did not correspond with the greatest skinfold values measured. Differences were significant at all sites except the iliac crest and the thigh ($P < 0.05$).

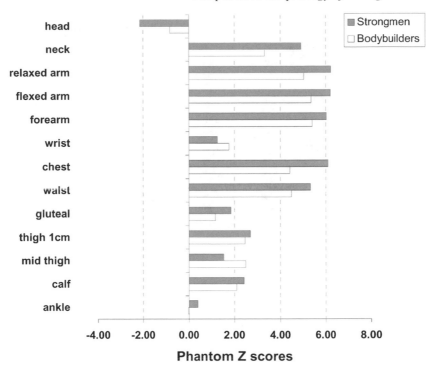

Figure 3a. Girth Profile

With the exception of head girth and ankle girth both groups displayed large positive values which were relatively similar. Strongmen had greater chest girth (P < 0.05) than bodybuilders. None of the other differences was significant.

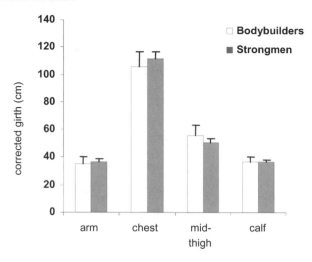

Figure 3b. Corrected Girth profile (height-adjusted)
Error bars refer to 1SD

Using the corrected-girth methodology, where the appropriate skinfold multiplied by *pi* is subtracted from the total girth, no significant differences were observed (P > 0.05). However, there was a tendency for bodybuilders to have greater thigh corrected girth (P = 0.08).

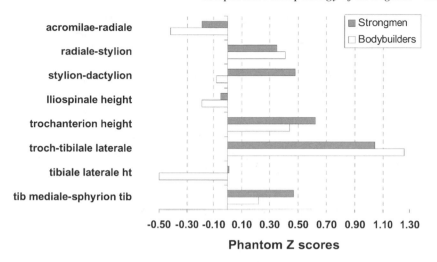

Figure 4. Segmental Length Profile

Differences in skeletal proportions were relatively minor between the groups and did not reach significance (P > 0.05).

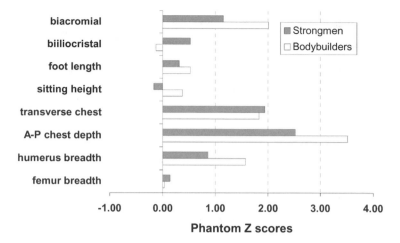

Figure 5. Bone Breadth Profile

Bodybuilders displayed significantly greater biacromial breadth (P < 0.05) and sitting height (P < 0.01) than strongmen.

The size adjustment calculation made for all variables normalised them to the phantom stature of 170.18 cm (Ross and Wilson, 1974). It was hypothesised that proportional skeletal dimensions, and corrected girths would be similar between the athletic groups, but that that skinfolds would be greater in the strongmen. The results showed the adjusted sum of 8 skinfolds to be greater in strongmen (than bodybuilders 159 ± 26 v 118 ± 32 mm P < 0.01), and the adjusted corrected girths were not significantly different, though the direction of the difference varied by region (e.g. corrected thigh, 50.8 ± 2.5 v 55.4 ± 7.1 P = 0.08; corrected arm 36.6 ± 2.2 v 35.4 ± 5.1 P = 0.52 for strongmen and bodybuilders respectively). Adjusted biacromial breadth (42.0 ± 2.1 v 40.3 ± 0.9 P = 0.03) and the androgyny index (biacromial breadth multiplied by 3, divided by bicristal breadth; 4.4 ± 0.2 v 4.1 ± 0.3 P < 0.01) and sitting height (91.6 ± 1.1 v 89.1 ± 1.9 P < 0.01) were greater in bodybuilders than strongmen.

4. DISCUSSION

Both athletic groups exhibited characteristics of large muscle mass, evidenced by the magnitude of girths and corrected girths. Strongmen, in addition to being heavier and taller, were fatter than bodybuilders. However, the bodybuilders were fatter than others reported previously (Borms et al., 1986; Spenst et al., 1993; Withers et al., 1997). This may be affected by the timing of the study to coincide with the maximum total mass, and presumably maximal fat mass of the bodybuilders, before the contest preparation had commenced. This was a deliberate strategy, because little is known about the maximum body mass and adiposity of bodybuilders at this stage, relative to the competition stage.

Comparison of the present study's bodybuilders showed identical total mass and mesomorphy scores to those of the heavy weight category bodybuilders of the study of Borms et al. (1986) measured within one to three days of the world amateur championships competition of 1981 in Cairo. However, mean endomorphy scores were dramatically lower in the latter (mean of 1.75 versus 4.1 in the present study). A logical explanation might be that the Cairo heavy weight bodybuilders had greater muscle mass than those of the present study, and had measurements been made two months beforehand, might have shown similar skinfold values. While it would be premature to imply causation with such comparisons of cross-sectional data, the spectacular difference in morphology associated with contest preparation of elite bodybuilders would appear to reduce dramatically the 'veneer' of adipose tissue through which the underlying muscles are seen. It is likely that in addition to exercise and nutritional regimes, their preparation in the days leading up to the contest included significant dehydration. At the time of the weigh-in, the subjects appeared to one of the measurers as in a state of 'reversible death' (Ross, 2005, personal correspondence). The same publication alludes to an observation of the present study, that strength gain is accompanied by fat gain, in addition to the desired muscle bulk. The heaviest category bodybuilders in Borms et al. (1986) were significantly higher in endomorphy than other groups, suggesting that some fat acquisition and retention may be a consequence of gaining a very large muscle mass. In the present study, the strongmen, whose muscle mass exceeded that of the bodybuilders, were also significantly fatter.

Spenst et al. (1993) examined the muscle mass of male athletes from several different sports including body building, sprinting, basketball and gymnastics. Bodybuilders and basketball players had similar total mass, but bodybuilders had greater muscle mass, despite being an average 16 cm shorter in stature. Adjusting for body mass by analysis of covariance produced higher values for sprinters than for bodybuilders, although it could be argued the adjustment should have been made for stature instead. However, bodybuilders in the present study were considerably fatter (averaging 123 mm for $\Sigma 8$ skinfolds, as compared to 62 mm for $\Sigma 9$ skinfolds). Comparing the two samples, the apparent reduction in fat mass is estimated to be from 16% to 8.8% using the equation of Stewart and Hannan (2000) which was developed for athletes. Athletes of the present study were a minimum of two months prior to competition, and were at their fattest and heaviest, about to commence their contest preparation. Bodybuilders in the study of Spenst et al. (1993) had competed within the previous 6 months, but the actual timing was not reported. It is likely that after competition, subjects retained a degree of leanness not shown by participants in the present study. A different study of body composition change over time before competition with three subjects using a four-compartment method, showed average $\Sigma 8$ skinfold totals of 51 mm 10 weeks prior to competition, reducing to 37 mm five days before the actual competition (Withers et al., 1997). Using the same % fat prediction (Stewart and

Hannan, 2000), the athletes were 6.9% fat 10 weeks and 5.1% fat 5 days prior to competition, respectively. Despite the low subject numbers, such athletes clearly display much greater leanness than those of the present study, even before they commence their final reducing phase of training. The weight loss during the pre-competition phase of 6.9 kg was 64% fat loss. An earlier study using a two compartment model on 26 bodybuilders reported a 6.4 kg pre-competition weight loss of which 76% was estimated to be fat (Goris and Somers, 1986).

However, most successful bodybuilders are associated with considerable upper body bulk, and optimised proportions such as the torso length/stature and biacromial/biiliocristal breadth ratios (Fry et al., 1991) which such athletes are not able to alter. Somatotype studies of bodybuilders all show extreme mesomorphy and low endomorphy, due to very large muscles being highly visible through a relatively thin veneer of overlying adipose tissue. Because height contributes negatively to the mesomorphy calculation, it is perhaps not surprising to observe elite competitors at having only average stature or less (Carter and Heath, 1990).

An important finding is the difference in size and skeletal shape between the athletic groups. While both groups appear robust and muscular, strongmen are clearly of large stature, which confers biomechanical advantage in pushing, pulling, throwing and certain lifting tasks, where, for instance, heavy objects must be dragged, lifted or thrown over a certain height. Although the precise tasks vary with each competition, one example is lifting a 100 kg concrete sphere onto a platform two metres in height, which is a disadvantage for competitors of lesser stature. By contrast, bodybuilders are of smaller stature, but have a larger biacromial : biiliocristal ratio, enabling a 'trapezoid' shape to prevail, previously reported to be a discriminator for success (Fry et al., 1991). Greater sitting height in bodybuilders after size adjustment (P < 0.01) may also reflect unusual skeletal proportions, but more probably reflects the greater relative leg-length associated with increased stature (Nevill et al., 2004).

The high competitive standard of bodybuilders of the present study suggests that morphological differences may be attributed to differences in contest preparation practices involving greater soft tissue loss than have been previously reported. Strongmen, by contrast, appear to require a large stature, in addition to specific muscular strength and additional fat mass does not appear to hinder this performance. Because strength training and pre-competition preparation is unlikely to alter skeletal proportions, it is likely that successful athletes self-select into strength sports in which they are likely to excel. As the future interest in strength sports is likely to increase, future research may involve areas beyond morphology including disciplines such as body image and functional comparison of both these groups and other strength athletes.

REFERENCES

Borms, J., Ross, W.D., Duquet, W. and Carter, J.E.L., 1986, In *Perspectives in Kinanthropometry*, J.A.P. Day (Ed). The 1984 Olympic Scientific Congress Proceedings Vol 1. Champaign, IL: Human Kinetics, pp. 81-90.

Carter, J.E.L. and Heath, B.H., 1990, Somatotyping: Development and applications. Cambridge, UK, Cambridge University Press.

Duquet, W. and Carter, J.E.L., 2001, In *Kinanthropometry and Exercise Physiology Laboratory Manual volume 1: Anthropometry* (edited by R. Eston and T. Reilly), (London: Routledge), p. 55.

Fry, A.C., Ryan, A.J., Schwab, R.J., Powell, D.R. and Kraemer, W.J., 1991, Anthropometric characteristics of body-building success. *Journal of Sports Sciences* **9**, pp. 23-32.

Goris, M. and Somers, K., 1986, Body composition in male Belgian body builders in training and competition periods. In *Kinanthropometry III* (edited by T. Reilly, J. Watkins and J. Borms) (London: E.& F.N. Spon), p. 35.

Huygens, W., Claessens, A.L., Thomis, M., Van Langendonck, L., Peeters, M., Philippaerts, R., Meynaerts, E., Vlietinick, R. and Beunen, G., 2002, Body composition estimations by BIA versus anthropometric equations in body builders and other power athletes. *Journal of Sports Medicine and Physical Fitness* **42**, pp. 45-55.

International Society for the Advancement of Kinanthropometry, 2001. *International standards for anthropometric assessment.* Adelaide, SA.

Martin, A.D., Spenst, L.F., Drinkwater, D.T. and Clarys, J.P., 1990, Anthropometric estimation of muscle mass in men. *Medicine and Science in Sports & Exercise* **22**, pp. 729-733.

Nevill, A.M. Stewart, A.D., Olds, T. and Holder, R., 2004, Are adult physiques geometrically similar? The dangers of allometric scaling using body mass power laws. *American Journal of Physical Anthropology* **124**, pp. 177-182.

Ross, W.D. and Wilson, N.C., 1974, A stratagem for proportional growth assessment. *Acta Paediatrica Belgica*, Suppl. **28**, pp. 169 – 182.

Ross, W.D., 2005, personal correspondence.

Spenst, L.F., Martin, A.D. and Drinkwater, D.T., 1993, Muscle mass of competitive male athletes. *Journal of Sports Sciences,* **11**,pp. 3–8.

Stewart, A.D. and Hannan, W.J., 2000, Body composition prediction in male athletes using dual X-ray absorptiometry as the reference method *Journal of Sports Sciences* **18**, pp. 263-274.

Withers, R.T, Noell, C.J., Whittingham, N.O., Schultz, C.G. and Keeves, J.P., 1997, Body composition changes in elite male bodybuilders during preparation for competition. *The Australian Journal of Science and Medicine in Sport*, **29**, pp. 11 – 16.

CHAPTER THREE

Built for Success: Homogeneity in Elite Athlete Morphology

Timothy R Ackland
School of Human Movement & Exercise Science, University of Western Australia

1. INTRODUCTION

Athletes who compete in international competition generally display distinctive body size, shape and composition characteristics compared with the normal population. These characteristics often give them a significant competitive advantage. If the sub-population of elite performers does not conform to a distinctive physical structure, an advantageous morphology may be identified if the finalists or medalists at international competition (Best) can be shown to differ from the remaining competitors (Rest).

When seeking explanations for improved performances, too often the focus is directed toward the influence of technical or equipment advances. However, the evolution of elite performers also played a significant role in the establishment of recent sporting records (Norton and Olds, 1996).

Success in international competition is, nevertheless, multifactorial. Not all sports require athletes to conform to, or be in possession of, selected anthropometric traits. In fact, the coach who wishes to implement a talent identification strategy may be misguided when selecting persons who conform to average data, even if these were collected on elite performers. Therefore, the aim of this paper is to seek evidence for the existence of unique morphological characteristics among elite competitors across a number of sports. This will assist coaches to assess whether unique morphology characteristics provide a competitive advantage, and hence, might be included in a talent selection process.

2. THE TALENT IDENTIFICATION PROCESS

The model now adopted by many sports institutes around the world for the identification and development of athletic talent, and the preparation of elite performers for competition was published in 1979 by John Bloomfield. These ideas were further expanded in his book - Applied Anatomy and Biomechanics in Sport (Bloomfield et al., 2003).

This methodology has been used to good effect, especially in smaller countries like Australia, who regularly 'fight above their weight category' in the international sports arena. For example, with a population of only 20 million, Australia ranked fourth in the medal tally in Sydney, and was a clear winner on a per capita basis. Much of this success can be attributed to the roles of sports scientists and coaches within the Australian sports institutes and their adoption of the athlete profiling model.

This method is commonly referred to as 'The Talent Identification Process', but may also be used to monitor the status of athletes throughout the training cycle. The process usually involves five stages, though this paper will focus on the first and third aspects only:

- Understanding the important aspects for success in competition

- Recording a set of data on an athlete

- Gathering a set of normative or comparative data

- Using these data to construct a profile of the athlete

- Interpreting the profile to guide the selection process, or provide the basis for an ongoing training program.

3. THE IMPORTANT ASPECTS FOR SUCCESS

Our understanding of the important attributes for success leads us to ask whether athletes in a particular sport demonstrate distinctiveness in physical capacities. Distinctiveness in human morphology (being just one component of physical capacity), that could lead to a competitive advantage within an elite group of athletes, may be demonstrated by:

- homogeneity of physical structure among elite competitors;

- possession of unique physical characteristics not commonly observed in the normal population; and/or

- significant differences between the best athletes and lower level competitors (Ackland *et al.*, 2003).

3.1 Homogeneity of Physical Structure

With regard to this factor, we may consider the characteristics of the distribution – skewness and kurtosis, or perhaps the somatotype attitudinal mean, but in a recent paper, Ackland *et al.* (2003) demonstrated this concept by graphing the standard deviation (SD) within the sample for a group of Sprint canoe and kayak paddlers, with respect to other elite athletes (Table 1).

Further observations can be made for other athlete groups in Table 1. Clearly, the lightweight rower's body mass is constrained by the rules, so we would expect a low SD value. But similarly, low SD for rugby front row players suggests the mechanics of scrimmaging technique and/or the constraints of the event play a role in the self-selection of elite competitors. Furthermore, moderately high SD values for rugby players in positions other than the front row, mean that weight is not a discriminatory variable and point to the wider range of physiques that may achieve success in this sport. In contrast, high SD values were reported

Table 1. Sample standard deviation for body mass (kg) of elite male athletes

Low (<4 kg)	Moderate (4 – 9 kg)	High (9 – 13 kg)	Very High (>13 kg)
Rowing (light)	Rowing (open)	Cycling	Weight lifting
Volleyball	Swimming	Boxing	Wrestling
Rugby (front row)	Rugby (other)	Judo	Track & field
Canoeing	Field hockey		
Triathlon	Water polo		
	Diving		

for the weightlifting, wrestling and track & field groups in this table, since these athletes were not stratified by weight category or event.

In some sports, the requirement to fit into a team structure so as not to unbalance or otherwise hinder the performance of other team members, might also result in similarity in physical structure. There exist many examples of this form of optimization from a number of sports, including the prop forwards within a rugby scrum, members of a rowing eight or a sprint kayak team, cyclists in the team pursuit event, and members of a synchronized swimming team.

3.2 Possession of a Unique Physical Structure

The possession of a unique physical structure that might give an athlete a mechanical advantage in technique becomes a critical Talent ID parameter. This uniqueness can be illustrated using the Overlap Zone concept, as reported by Norton and Olds (1998). The Univariate Overlap Zone can be used if we are dealing with a single variable such as stature. It gives us the probability of someone from one population (for example the normal population) falling within another population (ie. elite competitors within a sport), and is affected by both the mean and variance of the two curves.

When dealing with two characteristics simultaneously, such as height and body mass, this probability can be illustrated using the Bivariate Overlap Zone, once again demonstrated by Norton and Olds (1998). In their example, the probability of someone from the normal male population (displayed as a series of density ellipses) matching the physical size of NBA players is shown for each decade from the 1940s through to the 1990s. Despite the secular trend in the population toward taller and heavier individuals (shifting the density ellipses upward and to the right), this had not kept pace with the distinctive characteristics of NBA players (also shifting upward and to the right).

Uniqueness in physical structure need not be confined to physical size parameters. Possession of a unique proportionality characteristic may equally provide an advantage for an athlete in some events. This point is well illustrated using brachial index scores for sprint and slalom paddlers from the Sydney Olympics (Ackland *et al.*, 2003) as well as the normative data presented in Norton and Olds (1996). The brachial index expresses the ratio of radius length to humerus length and directly affects the mechanical advantage of the athletes' upper limb for generating force, velocity and power. Elite paddlers have a particularly high brachial index (average = 79). This unique characteristic for paddlers sets them apart from other strength and power athletes (weightlifting and wrestling average = 72) as well as the normal population (average = 75).

Other size and proportionality characteristics were identified by Bloomfield *et al.* (2003) as providing a mechanical advantage for the competitor (Table 2).

Table 2. Unique proportionality characteristics among elite athletes

Characteristic	Low	High
Stature	Gymnasts Divers	Middle distance runners Jumpers Throwers Sprint swimmers
Relative lower limb length	Sprint runners Gymnasts Divers Wrestlers Weightlifters	Middle distance runners Jumpers Basketball players Volleyball players
Crural Index [1]	Long distance runners Gymnasts Heavyweight wrestlers Weightlifters	Jumpers Sprint swimmers Basketball players Volleyball players
Brachial Index [2]	Weightlifters	Throwers Sprint swimmers

Note 1. Crural Index = tibiale laterale height / thigh length · 100
   2. Brachial Index = radiale-stylion length / acromiale-radiale length · 100

3.3 Best versus Rest Analysis

When the group of high level athletes does not exhibit homogeneity in physical structure, an advantageous morphology can be indicated if the best performers differ from the remaining competitors. Mazza *et al.* (1994) performed such an analysis for breaststroke swimmers competing in the 1991 World Championships. In this example, male swimmers with a final ranking in the top 12 (n = 21) were compared to the remaining competitors (n = 18). The Best performers were clearly taller (F = 11.26; p < 0.002) and heavier (F = 7.82; p < 0.08), with significant differences observed in eight girths, four breadths, and eight upper and lower limb segment lengths compared to those in the Rest group. Thus, it appears that a muscular physique, and its implication for power development, is a discriminating factor in male breaststroke performance.

4. CHARACTERISTICS OF QUALITY NORMATIVE DATA

Having identified which variables are important for success, and with a set of measurements recorded for your athlete, the next task is to gather a sample of normative data for comparison. To be effective, these normative data must be of a certain standard, otherwise you may not expose the critical variables to work on with your athlete. Good quality normative data must be derived from high performance athletes, measured in recent years and appropriately stratified. Several examples are provided below to reinforce this concept. Although the examples are confined to simple variables of stature, body mass and adiposity, the influence on other discriminatory parameters is equally valid.

Quality data on high performance athletes are difficult to collect. They are often jealously guarded by sports institutes and professional teams, and the levels of bureaucracy one must work through to gain access to athletes at international competition is mind bending. Nevertheless, in this paper I will refer to studies where data were collected at Olympic Games and World Championships.

Normative data will rapidly diminish in utility when the sport experiences substantial changes in rules, equipment, and athlete preparation. These changes can affect the physical and physiological capacities that provide an advantage for success in the sport or event. Therefore, coaches should always seek to use the most up-to-date set of comparison data in order to create an individual profile.

The data must also accurately reflect the status of the athlete, so must be appropriate for the individual's age, gender, and category (player position, event, or weight category). Extreme caution should be exercised when trusting the average values reported in the literature. Too often, we rely on convenient data without regard to its impact on the resulting profile chart. Using average data for comparison can be very misleading, especially if the sample is:

- sub-elite or relatively dated;

- heterogeneous in quality;

- heterogeneous in morphology/capacity;

- comprised of varying player positions/roles;

- comprised of varying events within the sport.

4.1 Sub-elite or Relatively Dated Data

The OZ2000 project (Ackland *et al.*, 2001) recorded 400 sets of morphology and equipment data across three sports (rowing, sprint and slalom canoeing) at the Sydney Olympic Games. Following data collection of sprint paddlers (n = 23 female, 60 male; with 59% ranked in the top 10 competitors), we were hard pressed to find useful comparison data in order to give meaningful feedback to athletes. Not only were the published research quite dated (de Garay *et al.*, 1974; Carter, 1982), the recent samples included several sub-elite athletes at National level competition (Vaccaro *et al.*, 1984). Therefore, average differences of up to 5 cm in stature and 5 kg in body mass were found between the Sydney athletes and previously reported data.

4.2 Heterogeneous in Quality

At the Olympic Games, only one team from each country may compete. This often means that some of the world's best athletes will not attend the meet, and that a number of athletes who only just make the qualifying criteria are included in competition. Thus one can get quite a mix of performance quality even within an Olympic sample. For example, among open male rowers at Sydney, the finalists were, on average, 2.5 cm taller and 3.5 kg heavier than the remaining competitors (Ackland *et al.*, 2001). A similar problem exists in many World Championship events in such sports as Rugby, Cricket and Basketball to name only a few. Using the average values for the whole sample would prove quite misleading as a normative data set for talent development.

4.3 Heterogeneous in Morphology/Capacity

This is a similar concept to the previous one. However, even within an elite group (such as those at the World Championships events - where there is generally a higher quality of competitors that at the Olympics), there may exist significant differences in athlete morphology - caused by factors other than competition ranking. Triathlon provides a useful example since it comprises three disciplines (swimming, cycling and running) with the potential for three disparate body types being self-selected plus the opportunity for 'hybrid' body types.

Considering the open competitors from the Triathlon World Championships in Perth (Landers *et al.*, 2000), the variance in height (SD = 4.4 cm female, 5.9 cm male), body mass (SD = 4.7 kg female, 6.0 kg male) and adiposity (Sum8SF) in particular (SD = 13.4 mm female, 10.2 mm male), were unusually high for such an elite sample. On further analysis of the data (using factor analysis and regression techniques), it appeared that there were a variety of identifiable body types, including:

- swimming specialists who could run and cycle quite well, and

- running specialists who could swim and cycle quite well.

So, the attributes for success among the competitors were clearly different for each leg of the race.

4.4 Varying Player Positions/Roles

It comes as no surprise that, when dealing with sports with varying player roles, using the averaged normative data to create a profile would be quite misleading. We may consider data from the 1994 Women's World Basketball Championships (Ackland *et al.*, 1997) to illustrate this point.

Unless we stratify according to player position, the data cannot be used effectively to build a profile chart. Here, the considerable differences between guards, forwards and centres are clearly shown in Figure 1 for stature, body mass and adiposity (sum of eight skinfolds).

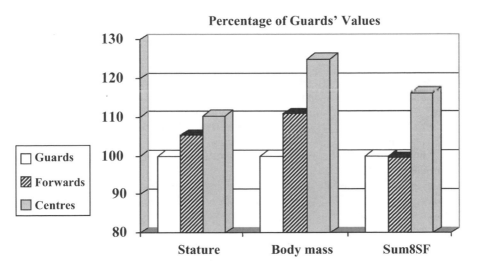

Figure 1 Body size and composition of World Championship women basketball players as a percentage of the score for guards

4.5 Varying Events within the Sport

Finally, when dealing with sports that have many sub-events such as swimming and track & field, using the averaged normative data to create a profile would also be quite misleading. To demonstrate this point, we may draw on data from the 1991 World Swimming Championships (Carter and Ackland, 1994), in which some 400 swimmers were measured.

Many research papers lump all athletes together for analysis. But, unless we stratify the data according to event (or similar events), the data may not be truly representative. In Table 3, the considerable differences between swimmers in each of these events are clearly shown (as a percentage of the value for freestyle swimmers) for stature, body mass, sum of skinfolds, and the percentage of muscle mass (estimated using the equations of Martin *et al.*, 1990).

Table 3 Morphology of specialist elite male swimmers expressed as a percentage of the value for freestyle stroke swimmers

Variable	Height	Mass	Sum6SF	% Muscle
Long distance (open water)	93.6	96.1	143.8	90.0
Breaststroke	97.8	94.6	92.4	103.0
Backstroke	99.8	97.5	94.5	100.0
Butterfly	97.8	95.3	97.8	101.0

5. CONCLUSION

A widely-employed model, first proposed by Bloomfield (1979), for the identification and development of athletic talent and the preparation of elite performers for competition was examined. Unfortunately, many coaches and sport scientists have not always used the process effectively. Often we see a huge battery of tests administered in a talent identification program, without too much forethought given to the importance of the selected parameters and use of quality normative data for comparison.

In simple terms, the main thrust of this paper is that one may be better off measuring fewer, but more discriminatory physical, physiological and psychological variables when applying this model, rather than the typical 'shotgun approach'.

Good quality normative data for comparison are essential in those sports for which a certain morphology is critical for performance success. Since the spread of scores in these discriminatory variables is small at the elite level, this serves to emphasize the requirement for quality reference data.

REFERENCES

Ackland, T., Kerr, D. and Schreiner, A., 1997, Absolute size and proportionality characteristics of world championship female basketball players. *Journal of Sport Sciences*, **15**(5), pp 485-490.

Ackland, T., Kerr, D., Hume, P., Norton, K., Ridge, B., Clark, S., Broad, E. and Ross, W., 2001, Anthropometric normative data for Olympic rowers and paddlers. In *Proceedings of the 2001 Annual Scientific Conference of Sports Medicine Australia*, edited by Ackland, T. and Goodman, C., (Perth: SMA) pp 1-5.

Ackland, T., Ong, K., Kerr, D. and Ridge, B., 2003, Morphological characteristics of Olympic sprint canoe and kayak paddlers. *Journal of Science and Medicine in Sport*, **6**(3), pp 285-294.

Bloomfield, J., 1979, Modifying human physical capacities and technique to improve performance. *Sports Coach*, **3**, pp 19-25.

Bloomfield, J., Ackland, T. and Elliott, B., 2003, *Applied Anatomy and Biomechanics in Sport, CDrom edition* (Melbourne: Blackwell Science).

Carter, J.E.L., 1982, *Physical Structure of Olympic Athletes: Part 1. The Montreal Olympic Games Anthropological Project.* Medicine and Sport Vol 16 (Basel: Karger).

Carter, J.E.L. and Ackland, T., 1994, *Kinanthropometry in Aquatic Sports* (Champaign, Human Kinetics).

de Garay, A., Levine, L. and Carter, J.E.L., 1974, *Genetic and Anthropometric Studies of Olympic Athletes*, (New York: Academic Press).

Landers, G., Blanksby, B., Ackland, T. and Smith, D., 2000, Morphology and performance of world championship triathletes. *Annals of Human Biology*, **27**(4), pp 387-400.

Martin, A., Spenst, L., Drinkwater, D. and Clarys, J., 1990, Anthropometric estimation of muscle mass in men. *Medicine and Science in Sports and Exercise*, **22**(5), pp 729-733.

Mazza, J., Ackland, T., Bach, T. and Cosolito, P., 1994, Absolute body size. In *Kinanthropometry in Aquatic Sports,* edited by Carter, J.E.L. and Ackland, T. (Champaign, Human Kinetics), pp 15-54.

Norton, K. and Olds, T., 1996, *Anthropometrica* (Sydney: UNSW Press).

Norton K. and Olds, T., 1998, The evolution of the size and shape of athletes: causes and consequences. In *Kinanthropometry VI*, edited by Norton, K., Olds, T. and Dollman, J. (Adelaide:ISAK), pp 3-36.

Vaccaro, P., Gray, P.R., Clarke, D.H., and Morris, A.F., 1984, Physiological characteristics of World class white-water slalom paddlers. Research Quarterly for Exercise and Sport, **55**(2), pp 206-210.

CHAPTER FOUR

A kinanthropometric profile and morphological prediction functions of elite international male javelin throwers

Ankebé Kruger, J. Hans de Ridder, Heinrich W. Grobbelaar, Colette Underhay. School of Biokinetics, Recreation and Sport Science, University of the North-West (Potchefstroom Campus), Potchefstroom, Republic of South Africa

1. INTRODUCTION

As it is the aim of any elite sportsman to compete and win at the highest level (De Ridder and Peens, 2000), and it is a rarity in sport for successful participants not to demonstrate morphological similarities to their peers, the influence of body build, form and composition (morphology) on performance has been researched off and on since the fifth century B.C (De Ridder, 1993). However, it was not until the late 20th century that this type of research came into its own.

Studies regarding the morphological characteristics of elite sportsmen help to understand the morphological, biomechanical and physiological requirements of performance, the optimal physical requirements for successful participation and the selection criteria for the identification of talented youth sportsmen (De Ridder et al., 2000).

Various studies have already been conducted with regard to the morphology or body build of javelin throwers (Carter, 1970, 1984a, 1984b; De Garay et al., 1974; Housh et al., 1984; Ross & Ward, 1984). The results of these studies show that certain morphological characteristics can be associated with participation in this specific athletics event. Such studies of javelin throwers, however, were usually part of studies for the general athletic population, or studies in which the morphology of a group of javelin throwers was compared to that of participants of another sport. Furthermore, studies conducted within athletics should differentiate between track and field athletes, with the latter able to be further subdivided into either throwers, i.e. those involved in shot-put, hammer throw, discus and javelin, or jumpers, i.e. those involved in long jump, triple jump, high jump and pole vaulting. The implication of subcategorising is that the group size of subcategories – javelin throwers among others - is relatively small. It is for this reason that limited research findings regarding the morphology of elite javelin throwers, as opposed to non-elite, exist.

Although research has been done on javelin throwers in determining various individual morphological components (Carter et al, 1982; Carter, 1984a; De Garay et al., 1974; Sharma & Dixit, 1985; Withers et al., 1986), a comprehensive kinanthropometric profile of elite javelin throwers has not been reported in the literature. The aim of this study, therefore was firstly to compile a complete kinanthropometric profile of elite male javelin throwers and secondly to identify the specific morphological variables which predicated international performance success in this field of athletics.

2. METHOD

2.1 Subjects

A one-off cross-sectional design was used for this study. Nineteen (N = 19) elite male javelin throwers with an average age of 26.4 ± 4.54 years were measured. Although it was a relatively small group, it was an elite and exceptionally homogeneous one. The subjects were classified according to their world ranking position as elite. All the javelin throwers were ranked in the top 88 in the world and included the world's numbers 1, 2, 4, 6, 7 and 9.

2.2 Procedure

The anthropometric variables selected for this study were: body mass, stature, armspan, bi-acromial breadth, chest breadth, bi-iliocristal breadth, wrist breadth, ankle breadth, humerus breadth, femur breadth, anterior-posterior chest depth, triceps skinfold, subscapular skinfold, supraspinale skinfold, abdominal skinfold, front thigh skinfold, medial calf skinfold, flexed arm girth, forearm girth, mid-thigh girth, calf girth; acromiale-radiale length and radiale-stylion length.

A Harpenden skinfold calliper with a constant pressure of 10 g/mm^2, a stadiometer, an electronic scale, a Lufkin metal measuring tape, a segmometer, a bone calliper and a large sliding calliper were used to measure the variables.

The anthropometric measurements were measured according to the techniques described in Norton et al. (1996). The subjects were measured mainly in the anatomical position (Marfell-Jones, 1996), with measurements taken twice and the average of the two values calculated for further data processing.

Variables were measured during the javelin throwers' yearly training camps in March 2002 and April 2003 in Potchefstroom, South Africa. Demographic information was obtained by means of a questionnaire. The subjects were informed regarding the nature of the measurements and were free to withdraw at any stage.

2.3 Statistical analysis

The Statistica computer package (StatSoft Inc., 2000) was used for the data analysis. Descriptive statistics were used to calculate the relevant variables for this study for the group as a whole, as well as for the group of 6 javelin throwers who ranked among the top 10 in the world (group A), and the remaining group of 13 javelin throwers who ranked between 11 and 88 in the world (group B). The independent t-test was used to evaluate the differences between the two subgroups (A and B). Statistical significance was set at $p < .05$.

The sum of the six skinfolds (Σ 6 skinfolds), triceps, subscapular, supraspinale, abdominal, front thigh and medial calf, were calculated and used as an indication of subcutaneous fat. The formula of Withers et al. (1987) was applied to determine the percentage body fat of the javelin throwers:

$$\text{Body density} = 1.10326 - 0.00031 \,(\text{AGE}) - 0.0036 \,(\Sigma \text{ 6 skinfolds}) \tag{1}$$

Where: (Σ 6 skinfolds) – sum of triceps, subscapular, supraspinale, abdominal, front thigh & medial calf skinfold measures.

$$\text{Estimated fat percentage} = (495/\text{BodyDensity}) - 450 \tag{2}$$

The formulas of Martin et al. (1990) and Martin (1991) were used to calculate muscle mass and skeletal mass respectively:

$$\text{Estimated muscle mass (kg)} = (\text{STAT} \times (0.0553 \times \text{CMTG}^2 + 0.0987 \times \text{FG}^2 + 0.0331 \times \text{CCG}^2) - 2445)/1000 \tag{3}$$

Where: STAT = stature (cm);
  CMTG = corrected midthigh girth (cm)
    CMTG = MTG – Pi x FTSF (mm)/10; Pi = 3.1416
  FG = uncorrected forearm girth
  CCG = corrected calf girth (cm)
    CCG = CG – Pi x MCSF (mm)/10

$$\text{Estimated skeletal mass (kg)} = 0.00006 \times \text{STAT} \times (\text{FEMRB} + \text{HUMRB} + \text{WRSTB} + \text{ANKLB})^2 \tag{4}$$

Where: STAT = stature (cm);
  FEMRB = femur breadth (cm);
  HUMBR = humerus breadth (cm);
  WRSTB = wrist breadth (cm);
  ANKLB = ankle breadth (cm)

Heath-Carter somatotypes were calculated to the nearest 0.1 for each component by using the formulas in Carter and Heath (1990). The average somatotypes for the total group of javelin throwers, as well as the individual somatotype for each javelin thrower, were calculated as follows:

$$\text{Endormorphy} = -0.7182 + 0.1451 \,(\Sigma \text{ of skinfolds} \times 170.18/\text{STAT}) - 0.00068 \,(\Sigma \text{ of skinfolds} \times 170.18/\text{STAT})^2 + 0.0000014 \,(\Sigma \text{ of skinfolds} \times 170.18/\text{STAT})^3 \tag{5}$$

Where: Σ of skinfolds = sum of triceps, subscapular,
supraspinale skinfolds
STAT = stature (cm)

Mesomorphy = 0.858 (HUMB) + 0.601 (FEMB) + 0.188 (CAG) +
0.161 (CCG) – 0.131 (STAT) + 4.5   (6)

Where: HUMB = humerus breadth (cm)
FEMB = femur breadth (cm)
CAG = corrected arm girth; (flexed arm girth (cm) –
triceps skinfold (mm)/10)
CCG = corrected calf girth; (calf girth (cm) –
calf skinfold (mm)/10)
STAT = stature (cm)

Ectomorphy = (HWR x 0.732) – 28.58   (7)

Where: HWR = STAT/ $\sqrt[3]{MASS}$
STAT = stature (cm)
MASS = mass (kg)

Note: If: HWR < 40.75, but > 38.25, then
Ectomorphy = HWR x 0.463 – 17.63   (8)

Note: If: HWR ≤ 38.25, then
Ectomorphy = 0.1   (9)

A forward stepwise discriminant analysis was performed to determine which morphological variables distinguished the elite (among the top ten in the world) from the remainder of the international throwers (among the top 88 in the world) in order to be able to identify those morphological characteristics that make a javelin thrower elite. The javelin throwers were divided into two groups according to world rankings. Group A represented the 6 javelin throwers among the world's top ten (positions 1,2,4,6,7,9), and Group B represented the 13 javelin throwers between positions 11 and 88. The differentiation ability of the classification function was determined by making use of a classification matrix. By means of this classification matrix, the javelin throwers (Groups A and B) were classified back into their original groups and the percentage of the number of javelin throwers in Group A and Group B correctly classified was also determined.

3. RESULTS

The descriptive statistics for measured variables for the javelin throwers as a whole and within each of the two sub groups, Group A (n = 6) and Group B (n = 13), are shown in Tables 1, 2, 3 and 4.

Table 1: Descriptive statistics for the male javelin throwers: base measures and girths (N = 19)

Variables	Group	$\overline{X}$	S.D.	Min	Max	p
Age (years)	A	28.7	4.08	24.0	36.0	0.001*
	B	25.3	4.48	20.0	33.0	
	TOTAL	26.4	4.54	20.0	36.0	
Body mass (kg)	A	94.4	5.41	88.6	102.0	0.429
	B	98.1	7.20	82.5	109.2	
	TOTAL	97.0	6.77	82.5	109.2	
Stature (cm)	A	185.6	3.11	181.8	190.3	0.646
	B	188.4	5.43	174.5	195.5	
	TOTAL	187.5	4.92	174.5	195.5	
Armspan (cm)	A	192.7	5.25	185.5	199.4	0.796
	B	193.1	6.53	182.7	204.0	
	TOTAL	192.9	6.01	182.7	204.0	
Flexed arm girth (cm)	A	38.4	1.93	35.6	40.5	0.892
	B	38.3	2.13	32.0	40.7	
	TOTAL	38.3	2.02	32.0	40.7	
Forearm girth (cm)	A	31.5	2.17	28.5	34.9	0.525
	B	30.1	1.48	27.0	32.5	
	TOTAL	30.5	1.78	27.0	34.9	
Mid-thigh girth (cm)	A	59.0	2.16	56.2	63.2	0.042*
	B	57.2	1.44	55.5	59.2	
	TOTAL	58.4	2.10	55.5	63.2	
Calf girth (cm)	A	40.9	1.32	39.2	43.3	0.941
	B	40.8	1.41	37.9	42.7	
	TOTAL	40.8	1.35	37.9	43.3	

* = p < .05

From Table 1 we note that the elite group (Group A) was significantly (p < .05) older than the sub-elite group (Group B). Although the elite group was lighter, shorter and had a slightly smaller armspan than the sub-elite group, none of these differences were significant. The elite group had wider arm girths (flexed arm and forearm) as well as wider leg girths (mid-thigh and calf) than the sub-elite group, with the only significant difference (p < .05) found in the mid-thigh girth.

Table 2: Descriptive statistics for the male javelin throwers: lengths and breadths (N = 19)

Variables	Group	$\overline{X}$	S.D.	Min	Max	p
Acromial-radial length (cm)	A	34.9	2.45	31.3	38.0	0.379
	B	35.4	1.57	33.0	37.9	
	TOTAL	35.3	1.83	31.3	38.0	
Radial-stylion length (cm)	A	28.2	1.42	26.9	30.0	0.906
	B	28.9	1.41	26.7	31.5	
	TOTAL	28.7	1.42	26.7	31.5	
Bi-iliocristal breadth (cm)	A	31.6	1.69	29.2	33.7	0.857
	B	32.1	2.09	28.6	36.0	
	TOTAL	31.9	1.94	28.6	36.0	
Bi-acromial breadth (cm)	A	46.1	2.57	43.2	49.5	0.999
	B	45.3	2.30	39.9	48.3	
	TOTAL	45.5	2.35	39.9	49.5	
A-P chest depth (cm)	A	21.7	0.70	21.0	22.7	0.446
	B	21.9	1.85	19.5	26.7	
	TOTAL	21.9	1.56	19.5	26.7	
Chest breadth (cm)	A	33.6	1.12	32.4	35.0	0.041*
	B	33.9	2.11	30.8	37.4	
	TOTAL	33.8	1.83	30.8	37.4	
Wrist breadth (cm)	A	6.1	0.25	5.9	6.3	0.750
	B	6.0	0.20	5.8	6.3	
	TOTAL	6.0	0.21	5.8	6.3	
Ankle breadth (cm)	A	8.3	0.06	8.2	8.4	0.563
	B	8.0	0.48	7.2	8.7	
	TOTAL	8.1	0.41	7.2	8.7	
Femur breadth (cm)	A	10.2	0.24	9.7	10.4	0.859
	B	10.2	0.60	9.1	11.3	
	TOTAL	10.2	0.51	9.1	11.3	
Humerus breadth (cm)	A	7.3	0.39	6.5	7.6	0.226
	B	7.5	0.32	7.1	8.2	
	TOTAL	7.4	0.35	6.5	8.2	

* = p < .05

As shown in Table 2, the elite group of javelin throwers had slightly shorter arms than the sub-elite group, but neither the difference in the acromial-radial length, nor the difference in the radial-stylion length was significant. From Table 2 we also note that although the elite group had smaller hips (bi-iliocristale breadth) and smaller chests (chest breadth and A-P depth) than the sub-elite group they had broader shoulders (bi-acromiale breadth) with the only significant difference being in chest breadth ($p < .05$) between the two groups. The elite group had the larger breadths with regard to the wrist and the ankle and was smaller than the sub-elite group with regard to the humerus. The two groups had the same measurement (10.2 cm) for the femur. None of these differences were statistically significant.

Table 3: Descriptive statistics for the male javelin throwers: component morphology (N = 19)

Variables	Group	$\overline{X}$	S.D.	Min	Max	p
$\sum$ of six SF[a] (mm)	A	50.6	6.98	40.9	61.1	0.081
	B	71.1	26.80	46.5	133.0	
	TOTAL	64.6	24.24	40.9	133.0	
Fat mass (kg)	A	9.4	1.25	7.5	11.0	0.246
	B	12.6	4.62	8.9	23.6	
	TOTAL	11.6	4.12	7.5	23.6	
% Body fat	A	10.0	1.28	8.4	11.7	0.348
	B	12.7	4.05	9.2	21.9	
	TOTAL	11.9	3.62	8.4	21.9	
Muscle mass (kg)	A	54.0	4.00	49.8	60.0	0.952
	B	54.2	5.53	47.3	66.9	
	TOTAL	54.1	4.98	47.3	66.9	
% Muscle mass	A	56.9	4.01	49.7	61.2	0.425
	B	55.4	3.67	47.4	61.3	
	TOTAL	55.8	3.74	47.4	61.3	
Skeletal mass (kg)	A	11.4	0.30	11.1	11.8	0.724
	B	11.1	0.95	9.5	12.8	
	TOTAL	11.3	0.74	9.6	12.8	
% Skeletal mass	A	11.9	0.55	11.3	12.7	0.933
	B	11.5	0.53	10.6	12.3	
	TOTAL	11.7	0.55	10.6	12.7	

$\sum$ of six SF[a] = triceps, subscapular, supraspinale, abdominal, front thigh, medial calf

It is clear from Table 3, that elite javelin throwers in Group A, had a smaller sum of skinfolds ($\sum$ of six SF) and therefore a lower fat mass (kg) as well as a lower percentage body fat, than the javelin throwers in the sub-elite group (Group B). None of these differences was statistically significant ($p < .05$). With regard to the muscle mass and the percentage muscle mass, Group A had a slightly smaller muscle mass (0.2 kg) than Group B, but a slightly higher percentage muscle mass (1.5%) than the sub-elite group. Neither of these differences was significant. Group A had larger skeletal mass (0.3 kg) as well as a higher percentage skeletal mass than Group B, but as was the case with all the other variables in Table 3, these two variables were not statistically significant.

Table 4: Descriptive statistics for the male javelin throwers: somatotypes (N=19)

Variables	Group	$\overline{X}$	S.D.	Min	Max	p
Endomorphy	A	2.1	0.35	1.7	2.6	0.097
	B	2.7	0.90	1.8	4.6	
	TOTAL	2.5	0.80	1.7	4.6	
Mesomorphy	A	6.2	0.80	5.6	7.5	0.083
	B	5.7	0.69	4.7	6.8	
	TOTAL	5.9	0.74	4.7	7.5	
Ectomorphy	A	1.4	0.52	0.7	1.9	0.759
	B	1.4	0.53	0.8	2.3	
	TOTAL	1.4	0.52	0.7	2.3	
S^b	A	2.1-6.2-1.4				
	B	2.7-5.7-1.4				
	TOTAL	2.5-5.9-1.4				

Sb Average somatotype of the group

As shown in Table 4, the average somatotype for the total group of javelin throwers was 2.5-5.9-1.4 and the total group can therefore be classified as endomorphic mesomorphs. Group A had a somatotype rating of 2.1-6.2-1.4 and Group B had a rating of 2.7-5.7-1.4 which classified both these groups also as endomorphic mesomorphs. The above-mentioned findings support the research results of Sharma and Dixit (1985) and Withers et al. (1986) where they found throwing athletes also to be endomorphic mesomorphs and this is similar to the findings of De Garay et al. (1974) who found that javelin throwers had a high mesomorphic value. An analysis of the individual components of somatotypes (endo-, meso- and ectomorph) showed that elite javelin throwers in group A had a higher mesomorphic component and a lower endomorphic component than the sub-elite javelin throwers in Group B, but neither the difference in the mesomorphic component nor the difference in the endomorphic component was significant. Group A and Group B had the same value (1.4) for the ectomorphic component.

Table 5: Forward stepwise discriminant analysis for the male javelin throwers (N = 19)

Step	Variable	F-value	p-value	WL
1	Age	10.53	0.008*	0.57
2	Body mass	3.97	0.072	0.39
3	% muscle mass	2.33	0.155	0.35
4	Mesomorphy	12.25	0.005*	0.61
5	Chest breadth	0.49	0.497	0.30
6	A-P chest depth	1.29	0.280	0.32
7	Armspan	5.44	0.040*	0.43

F (7,11) = 3,8515
WL = Wilks' lamda
* = p < .05

Table 5 represents a forward stepwise discriminant analysis that is used to determine which morphological variable best distinguishes between the javelin throwers in Group A and the rest of the javelin throwers in Group B. The discriminant analysis therefore shows which variables contribute to the fact that elite (top 10 world ranking) throwers can further be distinguished from other international throwers (top 88 world ranking).

Seven variables that maximally discriminate between the six javelin throwers among the top 10 (Group A) and the rest who rank among the top 88 (Group B) were identified by means of the discriminant analysis. The variables were age, body mass, % muscle mass, mesomorphy, A-P chest depth, chest breadth and armspan. According to the statistics in Table 5, age (F = 10.53, P < 0.05), body mass (F = 3.97, P < 0.05) and % muscle mass (F = 2.33, P < 0.05) best discriminate between the two groups. An important point to emphasize is that whereas age was an identifiable factor it is not supportable to identify it, in itself, as a cause of superior performance. It is, rather, an artefact of development and experience.

Table 6: The classification matrix for the male javelin throwers (N = 19).

Group	Percentage correct	Classification in groups	
		Group A	Group B
Group A (n = 6)	100%	6	0
Group B (n = 13)	92%	1	12

Table 6 represents the classification matrix of elite international javelin throwers. As shown in Table 6, 100% of the javelin throwers in Group A were classified back to their original group, and 92% of the remaining javelin throwers were classified back to their original group. One of the javelin throwers in Group B was classified back to Group A. The javelin throwers in Group A can therefore be distinguished from the other javelin throwers by virtue of the mentioned morphological variables (see Table 5).

It seems obvious that the group of elite javelin throwers is an exceptionally homogeneous group of athletes, distinguishable by the discriminant analysis even from the remaining group of javelin throwers that rank among the top 88 in the world.

4. DISCUSSION

The elite group (Group A) was significantly ($p < .05$) older than the sub-elite group (Group B). Although the elite group was lighter, shorter and had a slightly smaller armspan than the sub-elite group, none of these differences were significant. With regard to the girths (see Table 1) the elite group had wider arm girths (flexed arm and forearm) as well as wider girths with regard to the leg (mid-thigh and calf) than the sub-elite group, with the only significant difference ($p < .05$) in mid-thigh girth. The javelin throwers in this study were older, slightly taller (187.5 cm) and heavier (97.0 kg) than the javelin throwers in the study of Morrow et al. (1982). These results were further supported by the fact that the javelin throwers were taller and heavier than javelin throwers who took part in the 1960-1976 Olympic Games (Carter, 1984a) and taller and heavier than those throwers in the study of De Garay et al. (1974).

With regard to lengths and breadths, the elite group of javelin throwers had slightly shorter arms than the sub-elite group, but none of the arm lengths, were significant. From Table 2 we also note that although the elite group had smaller hips and a smaller chest than the sub-elite group they had broader shoulders with the only significant difference in chest breadth ($p < .05$) between the two groups. The elite group had the larger breadths with regard to the wrist and the ankle, was smaller than the sub-elite group with regard to the humerus and had the same measurement for the femur. None of these differences were statistically significant. In comparison with the study of De Garay et al. (1974), the javelin throwers of this study showed greater values for bi-acromiale (shoulder) breadth and bi-illiocristale (hip) breadth than the javelin throwers in the study concerned.

The elite javelin throwers in group A, had a smaller sum of skinfolds ($\sum$ of six SF) and therefore a lower fat mass (kg) as well as a lower percentage body fat, than the javelin throwers in the sub-elite group (Group B). None of these differences were statistically significant ($p < .05$). With regard to the muscle mass and the percentage muscle mass, Group A had a slightly smaller muscle mass (0.2 kg) than Group B, but a slightly higher percentage muscle mass (1.5%) than the sub-elite Group. Neither of these differences were significant. Group A had larger skeletal mass (0.3 kg) as well as a higher percentage skeletal mass than Group B, but as was the case with all the other variables in Table 3, these two variables were not statistically significant. The fact that current javelin throwers are heavier than in the past, as a result of muscle mass and skeletal mass that possibly increase as a result of resistance training may be ascribed to improved training methods and also to the evolution of body size and shape in athletes. Another possibility of morphological differences might be due to the change of centre of gravity in javelins. The fat percentage (11.9%) and fat mass (11.6 kg) of the javelin throwers are higher than the reported 8.48% and 7.65 kg for javelin throwers according to Morrow et al. (1982). Unfortunately, no other research results could be found with regard to muscle mass and skeletal mass of javelin throwers for a suitable comparison.

The average somatotype for the total group of javelin throwers was 2.5-5.9-1.4 and the total group could therefore be classified as endomorphic mesomorphs. Group A had a somatotype rating of 2.1-6.2-1.4 and Group B had a rating of 2.7-5.7-1.4 which classified both these groups also as endomorphic mesomorphs. The above-mentioned findings supported the research results of Sharma and Dixit (1985) and Withers et al. (1986) where they found throwing athletes also to be endomorphic mesomorphs. This is similar to the findings of De Garay et al. (1974) who found that javelin throwers had a high mesomorphic value. An analysis of the individual components of somatotypes (endo-, meso- and ectomorph) showed that elite javelin throwers in Group A had a higher mesomorphic component and a lower endomorphic component than the sub-elite javelin throwers in Group B, but neither the difference in the mesomorphic component nor the difference in the endomorphic component were significant. Group A and Group B had the same value (1.4) for the ectomorphic component.

5. CONCLUSION

The results of this study indicate that elite international male javelin throwers are an exceptionally homogenous group of athletes with regard to their morphological characteristics. They can be described as tall, heavy athletes with large percentages of muscle and skeleton mass. They also have broad, large chests with large shoulder and hip breadths and exceptionally long arms. They are classified as endomorphic mesomorphs. Although there were differences between the elite group (Group A) and the sub-elite group (Group B) of javelin throwers, only a few of these differences (age, mid-thigh girth and chest breadth), were statistically significant ($p < .05$). They are a group of athletes with a common morphology for their particular sport, with few significant morphological differences between the elite and the sub-elite throwers in the world. It is concluded that there has been a shift in the morphology of elite male javelin throwers over the past three decades, with these athletes in general becoming taller, heavier and more muscular than their predecessors. This parallels developments over the same period seen in other professional sports.

REFERENCES

Carter, J.E.L. (1970). The somatotypes of athletes – a review. Human biology, 42,535-569.

Carter, J.E.L. (1984a). Age and body size of Olympic athletes. In Physical structure of Olympic athletes. Part ll. Kinanthropometry of Olympic athletes (edited by J.E.L. Carter), pp. 53-79. Basel: Karger Press.

Carter, J.E.L. (1984b). Somatotypes of Olympic Athletes from 1948 to 1976. In Physical structure of Olympic athletes. Part ll. Kinanthropometry of Olympic athletes (edited by J.E.L. Carter), pp. 80-109. Basel: Karger Press.

Carter, J.E.L. and Heath, B.H. (1990). Somatotyping - development and applications. Cambridge, MA: Cambridge University Press.

Carter, J.E.L., Aubry, S.P. and Sleet, D.A. (1982). Somatotypes of Montreal Olympic athletes. In Physical structure of Olympic athletes. Part 1. Kinanthropometry of Olympic athletes (edited by J.E.L. Carter), pp. 53-80. Basel: Karger Press.

De Garay, A.L., Levine, L. and Carter, J.E.L. (1974). Genetic and anthropological studies of Olympic athletes. New York: Academic Press.

De Ridder, J.H. (1993). In Morfologiese profiel van junior en senior Cravenweek rugbyspelers. Potchefstroom: PU vir CHO (Proefskrif- Ph.D.)

De Ridder, J.H. and Peens, J. (2000). Morphological prediction functions for South African club championship cricket players. African journal for physical, health education, recreation and dance, 6,65-74.

De Ridder, J.H., Amusa, L.O., Monyeki, D., Toriola, A.L., Wekesa, M. and Carter, J.E.L. (2000). Kinanthropometry in African sports: Body composition and somatotypes of world class male African middle-, long distance and marathon runners. African journal for physical, health education, recreation and dance, 7,1-13.

Housh, T.J., Thorland, W.G., Johnson, G.O., Tharp, G.D. and Cisar, C.J. (1984). Anthropometric and body build variables as discriminators of event participation in elite adolescent male track and field athletes. Journal of sports sciences, 2,3-11

Marfell-Jones, M. (1996). Essential anatomy for anthropometrists. In Anthropometrica: A textbook of body measurement for sports and health courses (edited by K.L. Norton, & T.S. Olds), Marrickville, NSW: Southwood Press, pp. 3-24

Martin, A.D. (1991). Anthropometric assessment of bone mineral. In Anthropometric assessment of nutritional status (edited by J. Himes), New York: Wiley-Liss, pp. 185-196.

Martin, A.D., Spenst, L.F., Drinkwater, D.T. and Clarys, J.P. (1990). Anthropometric estimation of muscle mass in men. Medicine and science in sports and exercise, 22,729-733.

Morrow, J.R., Disch, J.G., Ward, P.E., Donovan, T.J., Katch, F.I., Katch, V.L., Weltman, A.L. and Tellez, T. (1982). Anthropometric, strength, and performance characteristics of American world class throwers. Journal of sports medicine, 22,73-79.

Norton, K., Olds, T., Olive, S. and Craig, N. (1996). Anthropometry and sports performance. In Anthropometrica: A textbook of body measurement for sports and health courses (edited by K.L. Norton, en T.S. Olds), pp. 289-352. Marrickville, NSW: Southwood Press.

Ross, W.D. and Ward, R. (1984). Proportionality of Olympic athletes. In Medicine Sport Science (edited by J.E.L. Carter), pp.110-143. Switzerland : Werner Druck AG.

Sharma, S.S. and Dixit, N.K. (1985). Somatotype of athletes and their performance. International journal of sports medicine, 6, 161-162.

StatSoft, Inc. (2000), STATISTICA (data analysis software system), version 6. www.statsoft.com

Withers, R.T., Craig, N.P. and Norton, K.I. (1986). Somatotypes of South Australian Male Athletes. Human biology, 58,337-356.

Withers, R.T., Graig, N.P., Bourdon, P.C. and Norton, K.I. (1987). Relative body fat and anthropometric prediction of body density of male athletes. European journal of applied physiology, 56,191-200.

CHAPTER FIVE

Athletic Morphology: Approaches and limitations using dual X-ray absorptiometry and anthropometry

Arthur D. Stewart

School of Health Sciences, The Robert Gordon University, Aberdeen
UK.

1. INTRODUCTION

Body morphology is a determinant of functional capacity. It is both a cause and a consequence of the relationship an organism has with its environment. Human body composition, together with biomechanics, psychology, physiology and the effect of the ambient environment, occupies an inter-dependent status with respect to exercise capability and sporting performance. The biomechanical imperative in different sports commonly involves optimising body size or mass, maximising force production and hence acceleration, or perhaps minimising energy expenditure. Athletic ability and training, from a physiological perspective can be seen to deliver these objectives. Although it is widely recognised that sporting success is multi-factorial, the hypothesis can be constructed that in sports with a clear biomechanical or indeed aesthetic mandate, the ultimate performance requires the ideal physique and body composition for that sport. Once it is clear that a sporting activity conforms to the hypothesis, the next objective is to define the nature of the ideal physique in terms of size, shape, composition, proportion and symmetry.

1.1 Physique

Physique refers to the phenotype or visible expression of the body structure. The study of physique has a rich history, articulated in various statues by the sculptors of Ancient Greece and later in the Renaissance period. In the early 20th century, interest in specific morphology readily identifiable by appearance became apparent in medicine where dysplasia was related to psychiatric and clinical disorders. "Athletic" accompanied "pyknic" (compact) and "asthenic" (lacking strength) as the three types of constitution (Kretchmer, 1936). To be athletic was viewed as a generic concept, conveying a suitability for several different types of activity. Based on such concepts further study identified a potentially infinite variety of physiques between pre-determined end points (Sheldon et al., 1940) and identifiable components of athleticism by rating of standard photographs. The study of physique is still useful today, when athletes may be third generation competitors. Some may have inherited favourable skeletal size and proportions for certain pursuits involving advantageous lever lengths and segmental inertia, and have self-selected into sports in which they are likely to excel. This concept is termed "*morphological optimisation*" (Norton and Olds, 1996). In addition to this, the conditioning programme (comprising periodised blocks of training) triggers variation in physique – specifically the adipose and muscle tissue masses - which is an anatomical expression of training, tapering and peaking for performance. This is referred to as "*morphological prototype*" (Hawes and Sovak, 1994). Because an adult is not at liberty to alter gross skeletal size or proportions, an understanding of these two concepts can, firstly, inform a prudent selection of sports in which to compete, and, secondly, explain how training may shape the physique on its skeletal template by adjusting muscle and fat mass relative to the proximity of competition.

1.2 Body Composition: the quest for a "Gold Standard"

Body composition refers to the structure behind the physique. As a scientific discipline within medicine, clinical nutrition and human physiology, it quantifies the proportions of the constituents which make up total mass. These constituents can be defined at the atomic, molecular, cellular or tissue level, and are expressed in absolute (mass) or relative (percentage of total mass) terms. Different methods of assessment will assume either a molecular model based on properties of chemical constituents (such as dual X-ray absorptiometry) or an anatomical one based on tissue masses or volumes (such as anthropometry). Combinations of models are possible, by dividing the body into hierarchical cascades of constituents.   This approach embraces all methods of body composition assessment, and has been central to the quest for the "gold standard" for using as a criterion method to represent best practice.

Historically, densitometry (where the body is subdivided into fat mass and fat-free mass, based on assumed densities of each) was viewed as such a criterion. More recently, the fat-free mass can be subdivided into bone mineral mass + lean soft tissue mass + total water content, which, along with fat mass, produce a four compartment (4C) model, providing a more accurate reference method. Predicting these masses in an athlete is possible to within a few percent of total mass, though the procedures are time consuming and expensive.   Alternatively, magnetic resonance imaging (MRI) could equally be argued to occupy "gold standard" status, as it provides a "photographic" dissection of the body which provides tissue-specific dimensions, which can be converted into masses using various density assumptions.   Although appealing, this method is very expensive, and usually limited to clinical applications. However, there is a more fundamental problem: against which absolute criterion are the 4C or MRI methods judged? The only possible answer is cadaver dissection and chemical analysis of the constituents. Without this, all methods are predictions, and, as a consequence, there is not one single living human for whom the body composition is accurately known.

Cadaver predictions can provide quantification of skeletal, muscle, adipose (+ skin) and residual masses, but are relatively rare (e.g. Drinkwater *et al.*, 1986).

Most predictions in use today only predict a single constituent, and make assumptions about the other constituents' proportions and density. Most are also doubly indirect, which means they predict an estimate of cadaver constituents. This is particularly the case in fat estimation via densitometry, where the assumed constancy of the density of the constituents of the fat-free mass has been shown to be false (Martin *et al.*, 1986), and the result is profound variations in predicted fat content due to invalid assumptions. Athletes present yet a further difficulty, because there are no published cadaver reports on dissected tissue masses of athletes, and it is clear that their body composition differs from those of other cadavers.

1.3 Level of fatness

In any individual, body composition reflects the balance of metabolic demands and energy supply. Throughout humankind's evolutionary past, survival has been contingent on a minimum "energy store" which would wax and wane with an intermittent food supply, in relation to energy expenditure. Considerable confusion surrounds the frequently overlooked distinction between *fat* (the mass of ether-extractable lipid in the body) and *adipose tissue* (the tissue containing the lipid, in addition to non-lipid constituents). The relationship between fat mass and adipose tissue mass is contingent on the *lipid fraction* (the fraction of the total cell or tissue mass which is fat). This shows variability intra- and inter-individually, and increases with adiposity, with the range estimated to be between 60 and 85% (Clarys *et al.*, 1987), displaying a strong positive correlation with total body fatness (Martin *et al.*, 1994). (Similarly, a lipid fraction applies to non-adipose tissue – to a much lesser extent). These and other limitations necessitate valid prediction equations for %fat to be sample-specific for level of adiposity, and consequently the use of so-called "generalised" equations for lean athletes is questionable.

In contrast to much of the population of the Western world, today's athletes are not excessively fat, and the pressure of competition leads many to seek to reduce their fat levels to minimum safe levels.

1.4 Performance and fat level

The late George Sheehan MD described runners who "looked well" as being "at least five pounds overweight". In doing so, he alluded to the fact that the morphology associated with health (or its appearance) is not necessarily that of optimal performance. This enlightened statement presents athletes with a conflicting message containing two inherent dangers: firstly of failing to achieve their performance potential, because they are more concerned with their health, and secondly (and more commonly) of neglecting long term health for the sake of short term performance. While science and medicine have long recognised the essential roles adipose tissue plays, for example in manufacturing hormones and storing vitamins, athletes or coaches who view fat mass as detrimental to performance may have been slow to recognise its importance for health.

Attempts to quantify ranges of %fat to safe or at risk zones have been made (Frisch and McArthur, 1974; Cantu and Micheli, 1991) for understandable reasons, but have largely failed to acknowledge that the tools available for measuring %fat would at best provide accuracy to within three percent (and commonly worse in lean athletes). More fundamentally, individuals may be highly variable in their health and performance at the same predicted %fat level. This may be because the partitioning between superficial and visceral fat depots is affected by age, sex, adiposity and physical activity (Stewart, 2003a), all of which could lead to erroneous predictions for athletes, depending on the prediction formula and body composition method used. There appears to be a lower limit for body fat – about 4-5% in males, and higher in females. When body fat is reduced to this level, the body will preferentially lose fat-free mass, as the body reduces its caloric requirement to maintain "metabolically expensive" tissue. If females reduce fat below a certain level (estimated at 12% and 17% by Cantu and Micheli (1991) and Frisch and McArthur (1974) respectively), reproductive function may be compromised. While short term perturbations below this threshold may be commonplace amongst elite athletes, they are not without their consequences. Several physiological processes including bone remodelling are implicated, and athletes and coaches need to be advised by appropriately specialised medical practitioners in this regard.

1.5 Muscle Mass

Muscle, by contrast to fat, follows the principle of specificity, in that training one muscle group, while requiring compensatory contractions in some other muscles, will not necessarily develop muscle in another body segment. Athletes adjust strength training protocols to maximise muscular strength, power or endurance, in accordance with their performance requirements. Their mandate is that muscle mass must be worth the energetic cost of carrying it, as additional muscle bulk not required will generally diminish a sporting performance. Paradoxically, the visibility of muscle in the physique is less to do with muscle size, than the thickness of the overlying adipose tissue. Training for strength sports – i.e. training designed to increase muscle mass, frequently produces modest gains in adipose tissue in addition, and strength sports with weight categories generally exhibit a gradient of adiposity, increasing with total mass.

While training can induce substantial muscle mass gains, inactivity results in its loss, especially the less-efficient fast twitch fibres. Consequently, athletes must focus on "building" and "maintaining" phases of training at different times, to end up with the appropriate physique for performance.

1.6 Developing a paradigm

The paradigm for viewing morphology in athletes must take account of the physique variables apparent in the phenotype, as well as the body composition as a whole. It involves genetically-inherited characteristics – most particularly the skeletal conformation, the surface and total adiposity, the specific nature of muscle development in accordance with the sporting requirements, in a physique which is inherently dynamic and unstable over time.

2. TECHNIQUES

2.1 Anthropometry

Anthropometry – the use of standardised surface measurements of the body can describe the phenotype in a quick and cost-effective manner, though a high skill level is required of the operator. Anthropometry can provide data on skinfold thickness, fat patterning, corrected girth (where the skinfold multiplied by *pi* is subtracted from segment girth) as a surrogate for muscularity, and indices of various skeletal proportions commonly expressed as percentages, such as the *skelic* (lower limb length / sitting height), *brachial* (forearm length / upper arm length), *crural* (lower leg length / thigh length) or *androgyny* (biacromial / biiliocristal breadth). Such measurements can be made in field settings and may reveal the inherited and training-induced advantages which top athletes have over their non-athletic peers (Stewart, 2001), and explain performance differences in "best versus rest" analyses.

The early work of Matiegka (1921) was possibly the first published study where body composition of athletes was compared with others. Using an anthropometric method, the data showed gymnastics instructors had an average of 45% greater muscle mass and 29% less fat mass when compared with apprentice butchers, barbers and blacksmiths. Today it is clear that athletes of different sports display different quantities of muscle, both in relative and absolute terms (Spenst *et al.*, 1993). Also, segregation of sports into athletic groupings in terms of endurance, strength, speed and team sports allows for larger data assemblies to be considered, with greater capacity for inference. Recently, corrected thigh girth displayed group-specific variation with total mass in a sample of 478 athletes and controls (Nevill *et al.*, 2004). Mean estimated muscle mass exponents ranged from 0.96 in endurance athletes to 1.49 in speed athletes, a finding attributable to the requirement for acceleration and deceleration, applying Newton's second law (Force equals mass × acceleration).

Although anthropometry is most widely used to estimate relative adiposity of athletes, it is also a convenient indicator of adipose tissue distribution. This falls largely under hormonal control, and altering the prevailing endocrine climate can have recognisable effects, for example in the case of Cushing's disease (caused by excess glucocorticoids), fat centralises to the torso. Observations of fat patterning by skinfold ratios such as the subscapular:triceps or the abdominal:medial calf exhibits significant variation by sex, adiposity, age and exercise (Stewart, 2003a). Visual comparison of such patterns are conveniently made via the skinfold map – a radial plot, where the radial distance from a central point represents different skinfold sites, as illustrated in Figure 1. In fatter individuals, these readily indicate relative centralisation on the torso sites, while in athletes who have far less total fat, the relative distribution favours limb sites (specifically the thigh). In senior athletes it is observed that the age-related centralisation is partly negated by preferential lipolysis of abdominal adipocytes through high activity levels over several years. The "fit-fat" distribution reveals low torso and arm fat relative to controls, and provides further evidence of the unsuitability of predicting % fat from generalised formulae, especially if lower body sites are not used (Mavroeidi and Stewart, 2003). Distinguishing, on skinfold evidence, athletes who are healthy and

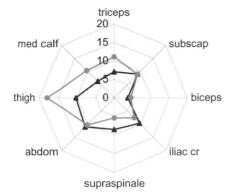

Typical male (dark) and female (light) profiles for healthy athletes across various sports. Average body fat values using dual X-ray absorptiometry were 8% and 14% for males and females respectively.

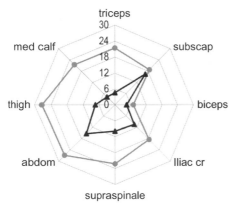

Profile of a male bodybuilder (dark) on anabolic steroids showing a large subscapular value and a female athlete (light) following recovery bulimia nervosa, showing relative uniformity of skinfold thickness.

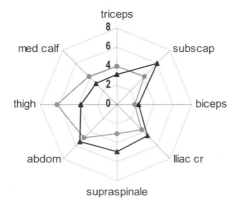

Very low skinfold totals in a male (dark) and female (light) endurance athletes both with compulsive exercise behaviour and Body Mass Index close to 16 kgm^{-2}. Note that the profiles converge to a considerable extent, and there is little evidence of sexual dimorphism.

Figure 1. Skinfold Maps of Male and Female athletes.
These are radial plots which follow the eight ISAK sites in a clockwise direction. Note each is drawn to a different scale.

extremely lean, from individuals suffering from disordered eating or compulsive exercise behaviour can be a more difficult task. While there are undoubted similarities between the conditions, where the sum of eight skinfolds may total less than 40 mm, chronically undernourished individuals are likely to exhibit signs of muscle wasting on the torso and arm, and possibly altered posture in addition.

2.2 Dual X-ray Absorptiometry (DXA)

Since the late 1980s, Dual X-ray Absorptiometry (DXA) has been the standard method for bone assessment. In addition to providing bone mass and areal bone density, it maps fat and lean soft tissue mass for each anatomical region, which can provide information on tissue distribution and relative symmetry (Stewart and Hannan, 2000a). The method has been validated on chemical analysis of pigs (Mitchell *et al.*, 1996) and in humans, there is evidence supporting DXA being more accurate as a reference method than densitometry (Prior et al., 1997). The technique has been used to validate skinfold-derived fat and fat-free mass in male athletes (Stewart and Hannan, 2000b).

Predictions of muscle mass from fat-free soft tissue also hold promise for athletes, although are confounded by limited reference data. Fat-free soft tissue includes all lean tissues of internal organs, blood and glycogen in addition to muscle. Predictions have centred on estimating total muscle mass from limb fat-free soft tissue, ignoring torso muscle, e.g. the prediction of Hansen *et al.* (1999) where total muscle mass is calculated as 1.33 times limb fat-free soft tissue mass, based on earlier work of very small numbers of cadavers. Such subjects may be morphologically distinct from athletes due to age and disease, and the specificity of training may further alter the muscle observed distribution in a sports-specific way, culminating in wide individual variation. Using a validated cadaver model, Stewart (2003b) found the co-efficients to average 1.41 and 1.24 for 106 male and 30 female athletes respectively across a variety of sports, suggesting that males not only have greater muscle mass than females, but a distribution which is more limb-based.

Body symmetry is another application conveniently assessed by DXA, and in uninjured athletes, bilateral fat free soft tissue mass differences have been observed to be as high as 24% in the arms of a badminton player and 8% in the legs of a fencer (Stewart, unpublished data). The ideal relative symmetry for an individual in a specific sport is unknown, but is a future research direction which may embrace body composition and the biomechanics of movement.

With all the convenience offered by DXA, it is not fully appreciated that its results are manufacturer-specific, and algorithms written to calculate soft tissue composition have not been designed to "expect" the morphology of athletes, which can lead to difficulty in interpreting results. As a criterion method, DXA is helpful, but caution is advised if athletic subjects are either excessively tall to fit into the scan area (approximately >190 cm), excessively thick (>25 cm) in an anterior-posterior plane, or be so undernourished that there are too few non-bone pixels for the system software to base soft tissue calculations from. This is illustrated in Figure 2, which illustrates data from a healthy and an undernourished individual. As a diagnostic tool, a DXA scan might be a result of a clinical referral for a sub-clinical stress fracture. However, in conjunction with anthropometry, the

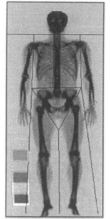

U1214950E   Thu 14.Dec.1995 16:32
Name: STEWART, Arthur
Comment:
I.D.:   R01A001   Sex: M
S.S.#:   - -   Ethnic: W
ZIPCode: Height: 177.00 cm
Scan Code: SC Weight: 70.80 kg
BirthDate: 27.Aug.58   Age: 37
Physician: STEWART
Image not for diagnostic use
   TBAR231
   F.S. 68.00% 0(10.00)%
   Head assumes 17.0% brain fat
   LBM 73.2% water

Region	Fat (grams)	Lean+BMC (grams)	% Fat (%)
L Arm	500.7	3194.8	13.6
R Arm	450.7	3385.6	11.7
Trunk	3753.2	30242.2	11.0
L Leg	1744.5	10242.1	14.6
R Leg	1753.6	10367.6	14.5
SubTot	8202.7	57431.5	12.5
Head	716.7	3808.0	15.8
TOTAL	8919.4	61239.5	12.7

·18.Dec.1995 14:27 [330 x 146]
Hologic QDR-1000/W (S/N 967 P)
Enhanced Whole Body V5.55

Scan output from a healthy individual showing regional fat distribution

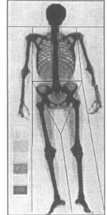

U0120980E   Wed 28.Jan.1998 14:53
Name:
Comment:
I.D.:   Sex: M
S.S.#:   Ethnic: W
ZIPCode: Height: 181.00 cm
Scan Code: Weight: 55.60 kg
BirthDate: Age: 21
Physician:
Image not for diagnostic use
   TBAR231
   F.S. 68.00% 0(10.00)%
   Head assumes 17.0% brain fat
   LBM 73.2% water

Region	Fat (grams)	Lean+BMC (grams)	% Fat (%)
L Arm	40.7	2853.4	1.4
R Arm	145.4	2935.8	4.7
Trunk	****	27612.4	0(3.7)
L Leg	140.9	7786.7	1.8
R Leg	153.4	8379.7	1.8
SubTot	****	50048.4	0(1.1)
Head	676.1	3823.9	15.0
TOTAL	141.0	54407.4	0.3

·29.Jan.1998 11:25 [330 x 146]
Hologic QDR-1000/W (S/N 967 P)
Enhanced Whole Body V5.67

Scan of an undernourished male (whose skinfolds are in Figure 1c). Asterisks refer to no detectable fat. Figures in brackets are negative predictions, based on absorptiometry algorithms which failed to detect sufficient non-bone pixels. In such individuals bone pixels may comprise up to 60% of the scanned area, and the assumed fat distribution model for basing the %fat predictions is invalid.

Figure 2. Dual X-ray absorptiometry whole body scans

utility of the fat prediction from the whole-body scan output might be enhanced. Experience with Hologic scanners (using enhanced whole body software versions 5.55 and 5.67) which remain the best-validated using athletic groups, has shown that when Σ8 skinfold total falls below about 40 mm, it is likely that DXA will fail to find any fat in some regions, and that interpretation of the scan data is necessarily guarded (Stewart, unpublished data). This might be pertinent in, for example, weight category sports. The pressure to perform drives many athletes to pursue not only exhaustive exercise regimes, but potentially unsafe practices for making weight. In other sports, the dimensions of the athlete may undermine DXA's ability to assess composition. Hence, groups such as basketball players, strongmen or Sumo wrestlers may not only be impractical to measure using DXA, but problematic to interpret results for.

3. DISCUSSION

3.1 The purpose of measuring body composition

With tools such as anthropometry and DXA, the utility of body composition measures in an applied setting lies in profiling change trajectories, with quantified precision error enabling the detection of meaningful differences. Here, the interests of the athlete, coach, scientist and clinician may converge, although each has a different perspective on the same picture. In optimising muscle, strength, speed and flexibility are pivotal. However in optimising fat, consideration needs to be given to balancing the short term performance with longer term health. In a "best practice" model, a consensus about strategies for testing will be the domain of national governing bodies of sport. Such testing will require the best possible data in terms of measurement protocols and quality assurance, with its inherent implication for robust training and examination procedures. Quality assurance for DXA is normally a matter for clinical audit, carried out by radiography or technical staff. For anthropometry, The International Society for the Advancement of Kinanthropometry (ISAK) is not the only organisational structure offering skills-based teaching in anthropometry. However, it remains the only protocol with quality assurance implicit in its procedures, so that error can be quantified in a site-specific and measurer-specific way. Such an approach not only makes the results more accurate, but also more useful.

3.2 The future of measuring body composition

Since Matiegka's early work, body composition assessment of athletes has enjoyed a rich history, but perhaps it can expect a more exciting future. It is likely that tomorrow's athletes will continue to be measured by anthropometry and DXA, whereas more sophisticated methods such as magnetic resonance imaging (MRI) and computed tomography (CT) with their excessive cost or unjustifiable X-ray exposure are likely to remain in the clinical domain. Of the other methods, the lower cost of ultrasound and 3D laser scanning, for rapid prototyping and change assessment offer considerable potential. The convergence of these technologies will unlock the door to multiple platforms of sports research, interfacing with dynamic measurements which are likely to transform our understanding of human morphology in sporting contexts (Olds, 2004). Armed with such information, tomorrow's sports scientists will be able to model theoretically perfect performances from "virtual athletes", and hopefully inform training-related decisions for maximising the potential in any real athlete's performance.

REFERENCES

Cantu, R.C. and Micheli, L.J., 1991, *American College of Sports Medicine: Guidelines for the Team Physician.* (Philadelphia: Lea & Febiger).

Clarys, J.P., Martin, A.D., Drinkwater, D.T. and Marfell-Jones, M.J., 1987, The skinfold: myth and reality. *Journal of Sports Sciences*, **5**, pp. 3-33.

Drinkwater, D.T., Martin, A.D., Ross, W.D. and Clarys, J.P., 1986, Validation by cadaver dissection of Matiegka's equations for the anthropometric estimation of anatomical body composition in adult humans. In *The 1984 Olympic Scientific Congress Proceedings: Perspectives in Kinanthropometry* edited by Day, J.A.P. (Champaign: Human Kinetics), pp. 221–227.

Frisch, R.E. and McArthur, J.W., 1974, Menstrual cycles: fatness as a determinant of minimum weight for height necessary for their maintenance or onset. *Science*, **185**, pp. 949–951.

Hansen, R.D., Raja, C., Aslani, A., Smith, R.C. and Allen, B.J., 1999, Determination of skeletal muscle and fat-free mass by nuclear and dual energy X-ray methods in men and women aged 51 to 84. *American Journal of Clinical Nutrition*, **70**, pp. 228-233.

Hawes, M.R. and Sovak, D., 1994, Morphological prototypes, assessment and change in elite athletes. *Journal of Sports Sciences*, **12**, pp. 235–242.

Kretchmer, E., 1936, *Physique and Character: An investigation into the nature of constitution and the theory of temperament.* 2nd ed. (London : Paul, Trench, Trubner & Co., Ltd).

Martin, A.D., Daniel, M.Z., Drinkwater, D.T. and Clarys, J.P., 1994, Adipose tissue density, estimated adipose lipid fraction and whole body adiposity in male cadavers. *International Journal of Obesity*, **18**, pp. 79-83.

Martin, A.D., Drinkwater, D.T., Clarys, J.P. and Ross, W.D., 1986, The inconstancy of the fat-free mass: a reappraisal with implications for densitometry. In *Kinanthropomtery III*, edited by Reilly, T., Watkins, J. and Borms, J. (London: E.& F.N. Spon), pp. 92-97.

Matiegka, J., 1921, The testing of physical efficiency. *American Journal of Physical Anthropology*, **4**, pp. 223–230.

Mavroeidi, A. and Stewart, A.D., 2003, Prediction of bone, lean and fat tissue mass using dial X-ray absorptiometry as the reference method. In *Kinanthropometry VIII*, edited by Reilly, T. and Marfell-Jones, M. (London: Routeledge), pp. 29–38.

Mitchell, A.D., Conway, J.M. and Potts, W.J.E., 1996, Body composition analysis of pigs by dual-energy X-ray absorptiometry. *Journal of Animal Science*, **74**, pp. 2663-2671.

Nevill, A.M., Stewart, A.D., Olds, T and Holder, R., 2004, Are adult physiques geometrically similar?: the dangers of allometric scaling using body mass power laws. *American Journal of Physical Anthropology*, **124**, pp. 177-182.

Norton, K. and Olds, T., 1996, *Anthropometrica*, edited by Norton, K. and Olds, T. (Sydney:. University of New South Wales Press), pp. 289–364.

Olds, T., 2004, The rise and fall of anthropometry. *Journal of Sports Sciences*, **22**, pp. 319-320.

Prior, B.M., Cureton, K.J., Modelsky, C.M., Evans, E.M., Slonogeer, M.A., Saunders, M. and Lewis, R.D., 1997, In vivo validation of whole body composition estimates from dual-energy X-ray absorptiometry. *Journal of Applied Physiology*, **83**, pp. 623–630.

Sheldon, W.H., Stevens, S.S. and Tucker, W.B., 1940, *The Varieties of Human Physique*. (New York: Harper & Brothers).

Spenst, L.F., Martin, A.D. and Drinkwater, D.T., 1993, Muscle mass of competitive male athletes. *Journal of Sports Sciences*, **11**, pp. 3–8.

Stewart, A.D., and Hannan, J., 2000a, Sub-regional tissue morphometry in male athletes and controls using DXA. *International Journal of Sport Nutrition and Exercise Metabolism*, **10**, pp. 157–169.

Stewart, A.D. and Hannan, W.J., 2000b, Body composition prediction in male athletes using dual X-ray absorptiometry as the reference method. *Journal of Sports Sciences*, **18**, pp. 263-274.

Stewart, A.D., 2001, Assessing Body Composition in Athletes. *Nutrition*, **17**, pp. 694-695.

Stewart, A.D., 2003a, Anthropometric fat patterning in male and female subjects. In *Kinanthropometry VIII*, edited by Reilly, T. and Marfell-Jones, M. (London: Routledge), pp. 195–201.

Stewart, A.D., 2003b, Mass fractionation in male and female athletes. In *Kinanthropometry VIII*, edited by Reilly, T. and Marfell-Jones, M. (London: Routledge), pp. 203–209.

CHAPTER SIX

Monitoring exercise-induced fluid losses by segmental bioelectrical impedance analysis

Alexander Stahn [1,2], Elmarie Terblanche [3] and Günther Strobel [1]

[1] Institute of Sports Medicine, University Hospital Charité, Campus Benjamin Franklin, Free University of Berlin, Germany
[2] Department of Medical Physiology, Faculty of Health Sciences, University of Stellenbosch, South Africa
[3] Department of Sport Science, University of Stellenbosch, South Africa

1. INTRODUCTION

Heavy exertion and severe bouts of exercise may cause substantial losses of body water. A number of studies have shown that hypohydration of slightly more than 1% of total body weight may already impair physical and mental performance (Cian et al., 2001; Sawka, 1992). Water losses in excess of 3% are regarded as potentially harmful. Extreme fluid losses can even cause failure of the cardiovascular system (Greenleaf, 1992). Hence, an accurate, simple and non-invasive technology to monitor hydration changes is particularly important for maintaining exercise performance, as well as general health.

The most direct way to determine total body water as well as intra- and extracellular water is through isotopic dilution techniques. Given the fact that three to four hours have to be allowed for the distribution of the isotope of water throughout the body, this method is not suitable for measuring fluid losses and shifts which are induced by typical exercise sessions lasting between twenty minutes and two hours. Though the time interval between two consecutive measurements could be reduced by increasing the concentration of the radioactive marker, this procedure is difficult to justify in the absence of clinical imperative.

A method that might overcome the drawbacks of isotope dilution is bioelectrical impedance analysis (BIA). Present BIA instruments use a tetrapolar electrode arrangement to apply a low, imperceptible alternating current (200–800 μA) via two outer (distal) electrodes while the inner (proximal) electrodes are used to measure the voltage drop across the body (usually between the right wrist and ankle). Current and voltage can then be used to calculate electrical impedance (voltage/current). Impedance (Z) is the frequency-dependent opposition of a conductor against an alternating current. Since impedance is inversely related to volume, the latter can be derived from the following relationship assuming an isotropic conductor with a uniform cross-sectional area:

$$V(cm^3) = \rho(\Omega.cm) \times \frac{l(cm)^2}{Z(\Omega)}$$

(1.1)

where   V = volume
        ρ = resistivity
        l = length
        Z = impedance

Most BIA-models employ a so-called whole-body approach in which the body is modeled as one single cylinder and electrodes are placed at the wrist and ankle of one side of the body (Figure 1.1). Usually, conductive length is replaced by height because it is easier to obtain than the actual length of the conductor (sum of arm, trunk and leg length).

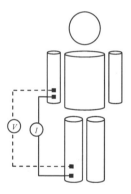

Figure 1.1 Whole-body BIA

Strictly speaking, resistivity should be replaced by impedivity in equation 1.1 if impedance instead of resistance is used to predict volume. The difference between impedance and resistance in biological conductors is, however, small (<10%) which is also the reason why the two terms have often been used interchangeably in BIA. Regardless, the BIA instrument used in the present study was not a phase-sensitive device and thus only allowed the determination of impedance. Furthermore, it is worth noting there is presently only one approach based on mixture theory from emulsion science which, in fact, employs resistivities for the calculation of fluid volumes (De Lorenzo et al., 1997a). In contrast, the majority of BIA-models are based on the impedance index (l^2/Z) and additional predictors such as weight, age and gender and use multiple regression analysis to derive their respective weights (regression coefficients) in validation studies in order to account for any violations and simplifications of the assumptions underlying Equation (1.1) (i.e. assumption of an isotropic cylindrical conductor of homogenous material with constant cross-sectional area and uniform current distribution).

In biological conductors current can be transmitted by both ionic and non-ionic issues. Most of the current is transmitted by ions dissolved in solution (intra- and extracellular water). The opposition of these electrolytes against the current is termed resistance (R) and comprises $\approx$90–98% of impedance in biological systems. The remaining part of impedance is attributed to reactance (Xc). Reactance is the inverse of capacitance and defines the amount of charge that can be stored by a capacitor. The vector sum of resistance and reactance describes impedance ($Z = R^{0.5} + X_c^{0.5}$). In the human body reactance can be mostly attributed to the lipid bilayer of cell membranes which act as capacitors when an alternating current is applied to the body (Figure 1.2).

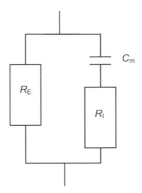

Figure 1.2 Simplified equivalent electrical circuit describing any biological impedance. R_E, extracellular resistance (Ω); R_I, intracellular resistance (Ω); C_m, capacitance of cell membranes (F).

Since the flow of charges "through" a capacitor is directly proportional to the rate of change of voltage across the capacitor, cell membranes behave like insulators at low frequencies. Hence, at frequencies close to zero, the current is determined by the extracellular fluid volume. At 50 kHz, the most commonly used frequency by commercially available BIA-instruments, some capacitive properties of the human body will be charged, allowing some of the current to pass through the intracellular space. Due to differences in cell membranes the amount of intracellular current pathway shows considerable variation between and within subjects when impedance measurements at a predetermined fixed frequency (e.g., 50 kHz) are employed. This is also reflected by the characteristic frequency, the frequency where reactance reaches a maximum, which is typically in the range of 35–85 kHz (Cornish *et al.*, 1996). At infinite, or at least at a very high, frequency there is effectively no build-up of charge on the cell membrane, and therefore no opposition on the part of the cell membranes to the current flow. As a result, at high frequencies, impedance is determined by both the extra- and intracellular fluids which allows the measurement of total body water (TBW) whereas at a frequency close to zero the current does not penetrate any cell membranes permitting the estimation of extracellular water (ECW) only. If reactance is plotted versus resistance this behavior is reflected by a semi-circle with a depressed center below the x-axis and typically known as a Cole-Cole plot (Figure 1.3).

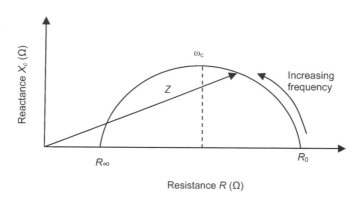

Figure 1.3 Schematic presentation of impedance locus. Reactance is plotted versus resistance for a range of frequencies. Z, impedance; R_0, resistance at 0 frequency (= extracellular resistance R_E); R_∞, resistance at infinite frequency (= extracellular and intracellular resistance R_I, i.e. $1/ R_\infty = 1/R_E + 1/R_I$); ω_c, characteristic frequency. At very low frequencies the impedance of cell membranes is high and thus blocks the conduction through the intracellular space. Hence, the measured impedance is purely resistive (R_0). As frequency increases the current penetrates the cell membranes and reactance also increases. Once frequency has reached a certain threshold, the characteristic frequency ω_c, where reactance is at a maximum, reactance decreases when frequency is further increased as cell membranes "lose" their capacitive characteristics. As a result, at high frequencies impedance is again purely resistive (R_∞). Furthermore, because the conducting volume increases with higher frequencies (increasing amount of intracellular fluid is conducting the current) measured impedance decreases (Z is inversely proportional to the volume of a conductor).

Though numerous studies have confirmed the validity of BIA to estimate TBW and ECW, a final conclusion if, and to what extent, BIA can be used for monitoring fluid shifts cannot be drawn as most validation studies have been conducted under standardized clinical conditions at normohydration. In spite of a promising trend to use BIA successfully in clinical settings — for example, BIA has been shown to track fluid changes during abdominal surgery, dialysis and pregnancy (Bracco *et al.*, 2000; Chamney *et al.*, 2002; Lozano *et al.*, 1995; Takeuchi *et al.*, 2000; Tatara and Tsuzaki, 1998; Zhu *et al.*, 2000) — results from studies in exercise physiology remain controversial. On the one hand, some investigators found BIA to predict TBW after exercised-induced hypohydration with acceptable accuracy (e.g., O'Brien *et al.*, 1999; Pialoux *et al.*, 2004). On the other hand, a number of studies failed to predict TBW after exercise (e.g., Collodel *et al.*, 1997; Koulmann *et al.*, 2000; Liang and Norris, 1993).

Whereas the above-mentioned discrepancy is likely to be due to different methodologies (i.e. time of measurement subsequent to exercise and type of recovery), all studies conducted on the effect of exercise on BIA have in common that they employed a conventional BIA approach. Conventional or whole-body BIA determines the voltage between the wrist and ankle of one (usually right) side of the body. This measurement configuration is based on the assumption that the human body is an isotropic conductor with homogeneous length and cross-sectional area. Obviously this assumption is incorrect as the human body is a complex of truncated cones, cylinders and ellipsoids of material which is anisotropic. A grave simplification of this complex is to model the human body as five cylinders, i.e. two arms and legs and a trunk with varying cross-sectional areas. Given the fact that impedance is inversely proportional to the cross-sectional area, the trunk contributes about 15% of total body impedance (Organ *et al.*, 1994) although it contains about 50% of the body mass (Clarys and Marfell-Jones, 1986). This means that even large variations of trunk impedance are not reflected in total body impedance whereas minor physiological changes in the limbs can cause a significant change in the overall impedance (Baumgartner *et al.*, 1989; Patterson *et al.*, 1988). Thus, a fluid loss in the trunk will hardly be detected by conventional BIA. Furthermore, since fluid shifts as a result of exercise can be limb-specific depending on the type of exercise performed (Sawka and Coyle, 1999) any estimation of body fluids by a whole-body measurement may be further confounded. Koulmann *et al.* (2000) noted that the influence of physical exercise might "[…] induce a relative increase of the state of hydration of those segments [lower limbs], which would partly conceal the decrease of fluid volumes occurring mainly in the trunk and the upper limbs". Finally, the fluid distribution between vascular, interstitial and intracellular compartments may be changed as a result of exercise (for review Senay, 1998). If BIA is performed at frequencies which do not entirely penetrate the cell membranes, any impedance measurements will only reflect a limited amount of total body water (Schoeller, 2000).

Impedance measurements at high frequencies (>100 kHz) and the separation of the body into five segments and their individual impedance measurements might alleviate the drawback of present BIA methodology. According to the authors' knowledge, no attempt has been made to estimate exercise-induced fluid changes by segmental impedance measurements of the arm, trunk and leg. Therefore the aim of the present study was to evaluate the efficacy of segmental BIA to determine exercise-induced losses of total body water. The primary focus of the study was to examine the relationships between impedance changes and changes of total body water following exercise using whole body and segmental BIA at high and low frequencies. The second aim of this study was to attempt to examine the effect of body temperature, blood chemistry and electrolytes on impedance. Finally, a reliability study was performed to assess the reproducibility of impedance measurements within the particular time frame of the present study.

2. METHODS

2.1 Sample

A total of 68 healthy Caucasian volunteers, 35 healthy women and 33 men, were invited to participate in the study. Subjects were recruited from local sports clubs via advertisements and from the University. Most of them were regularly engaging in physical activity several times a week (up to five times) during the previous six months. Exclusion criteria comprised any disorders (metabolic, cardiovascular, neurological or musculoskeletal) abnormalities or functional impairments that might either affect the outcome of the study or put the subjects at risk and/or use of any medication that may alter peripherical circulation. After a normal clinical investigation that included a detailed medical history, a physical examination and a general blood screening, subjects were randomly allocated to either the reliability or treatment (hypohydration) group in a ratio of 2:1. None of the subjects received any medication that was known to affect the variables measured. After the purpose, procedures, and known risks of the tests had been explained, written informed consent was obtained from all participants prior to the study, which was approved by the Ethics Committee of the Free University of Berlin.

2.1 Experimental design

Subjects of the reliability study were invited once for impedance measurements only. The hypohydration group was additionally asked to perform initially a maximal graded exercise test on the treadmill to determine $\dot{V}O_{2Max}$ on a separate day. All tests were performed in a thermally-controlled environment (20–22 °C ambient temperature and 55–60% relative humidity) at the Exercise Laboratory of the Institute of Sports Medicine, Free University of Berlin.

All subjects were asked to refrain from alcohol and heavy exertion or exercise for 48 h before trials and to refrain from smoking or drinking coffee on the day of the test. Moreover, they were advised to drink at least 2 L of water per day during the week preceding the trial to ensure normohydration. Additionally, they were instructed to have a light meal and drink a limited amount of water (not more than 1 L), 2 h pretrial to standardize hydration levels.

For the reliability study, an initial whole-body and segmental impedance measurement was performed after which subjects were asked to remain seated in a comfortable position for the following 60 min. Subsequently, measurements were repeated and finally performed again after sitting for an additional 60 min. Subjects were not allowed to consume fluid or food at any time during the treatment trial.

The hypohydration experiment consisted of the following baseline measurements that were obtained after voiding: weight, whole body and segmental impedances, skin surface temperatures, resting heart rate, blood pressure, hematocrit, and plasma sodium, chloride and potassium concentration. Subsequently, subjects exercised on the treadmill for 60 min (including a 5 min warm-up and cool-down phase, respectively) at a pace corresponding to 70–75% of their $\dot{V}O_{2Max}$. Immediately after termination of exercise, subjects were completely dried and changed their underwear and measurements were repeated. After resting in comfortable clothes in a seated position for an additional hour, the same measurements were taken again.

2.2 Testing

2.2.1 Direct Measurement of Maximal Oxygen Uptake

The test for the determination of $\dot{V}O_{2Max}$ was performed on a calibrated treadmill (XELG3 2616/93, Woodway, Weil am Rhein, Germany). The test protocol followed the guidelines provided by Lollgen *et al.* (1988). After a 5 min warm-up at a self-selected speed and a 1% gradient to simulate wind resistance, the speed was set at 2.0 m.s^{-1} for women and 2.5 m.s^{-1} for men and continuously increased by 0.5 m.s^{-1} every 3 min until subjects could not continue to exercise, despite verbal encouragement.

Pulmonary ventilation and gas exchange were measured with an open-circuit breath-by-breath online data acquisition system (OXYCONGAMMA®, Mijnhard, Bunnik, the Netherlands). The system was calibrated 1 hour prior to testing according to the specifications of the manufacturer. The mean oxygen uptake ($\dot{V}O_2$), carbon dioxide output ($\dot{V}CO_2$) and pulmonary ventilation ($\dot{V}E$) were computed for every breath and averaged over 10 s. Throughout the test, heart rate was continuously monitored with a heart rate monitor (PE 3000 Sports Tester®, Polar Electro, Kempele, Finland). 1, 3 and 5 min after termination of exercise a blood sample from a hyperemized (Finalgon Forte, Thomae, Biberach, Germany) ear lobe was collected in capillary tubes to determine the highest post-exercise blood lactate concentration (Ebio plus®, Eppendorf, Hamburg, Germany). The test was considered maximal if a leveling-off of $\dot{V}O_2$ was achieved (an increase of less than 150 ml.min^{-1} despite an increase in speed), or if at least two of the following criteria were met: (1) respiratory exchange ratio (RER) $\geq$ 1.15; (2) no less than 10 beats below age predicted maximal heart rate; (3) blood lactate concentration >8.0 mmol.L^{-1}.

2.2.2 Anthropometric Measurements

All anthropometric measurements were performed by two examiners using the "Anthropometric standardization reference manual" (Lohman *et al.*, 1988) as a guideline.

Body mass was measured barefoot, in minimal clothes on a calibrated electronic digital scale to the nearest 0.01 kg.

A stadiometer was used to measure standing height to the nearest 0.1 cm. Subsequently, arm, trunk and leg lengths were obtained for the calculation of segmental impedance indices. Measurements were taken in a standing position with arms and legs extended using a flexible steel tape (Lufkin, Cooper Tools, Apex, NC, USA). Each measurement was repeated and if the difference between the two values exceeded 0.5%, the procedure was repeated again. Whereas the center of the voltage sensing (proximal) electrodes served as a set of reference points for the measurements of leg and arm lengths, two additional reference points were defined as follows: Point A was defined as the center of a line between the *acromion* and the *axilla*, whereas point B is defined as the intersection of a horizontal line crossing the *anterior superior iliac spine* and a vertical line dissecting the thigh while subjects were standing. The distance between A and B is considered as the length of the trunk (i.e. the length of the conductive part allocated to the trunk) whereas the distances between A and the voltage sensing electrode of the hand and between B, and the voltage sensing electrode of the foot are the lengths of the arm and leg, respectively.

2.2.3 Bioelectric Impedance Measurement

Whole body and segmental impedance measurements were performed with a multifrequency bioelectrical impedance analyzer (Quadscan 4000®, Bodystat LTD, Douglas, Isle of Man, UK). Measurements were taken in triplicate and averaged from the right side of the body via a tetrapolar electrode arrangement following standard procedures (Lukaski *et al.*, 1985). Subjects, dressed in minimal clothes, were lying in a supine position with arms and legs 10° and 20° abducted from the body, respectively. Styrofoam positioning blocks were used to ensure correct posture throughout all measurements. Prior to electrode application, skin hair was removed, if necessary, and skin was rubbed with isopropanol (Softasept® N, Braun, Melsungen, Germany) for five seconds. The 16 cm² (8.0 cm × 2.0 cm) current introducing electrodes (Bodystat LTD, Douglas, Isle of Man, UK) were placed just below the phalangeal-metacarpal joint in the middle on the posterior surface of the hand and just below the transverse (metatarsal) arch on the superior side of the foot. Detector electrodes of the same type were positioned on the posterior surface of the wrist at the line bisecting the styloid processes of the ulna and radius, and in a line bisecting the medial and lateral malleoli and on the dorsal surface of the foot (Figure 1.4).

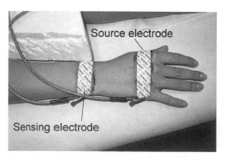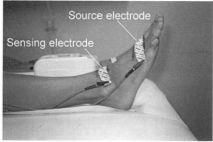

Figure 1.4 Electrode placement. Pictures taken from a female subject of the present study.

Two additional detector electrodes were placed at the equivalent positions of the contralateral side of the body to determine segmental impedance of the arms, legs, and trunk, respectively, as suggested by Organ *et al.* (1994). This measurement configuration neither requires the current injection electrodes to be moved nor to locate additional electrode sites other than the above-mentioned. To obtain separate impedance measurements of the right arm, leg and trunk the crocodile clips for measuring the potential difference were connected to the electrodes of the left and right wrists and ankles as shown in Figure 1.5.

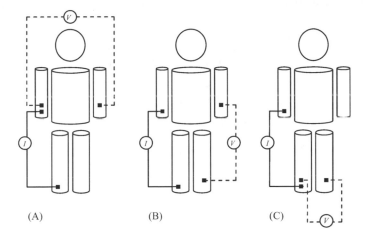

(A)   (B)   (C)

Figure 1.5 Segmental BIA. (A) Arm (B) Trunk (C) Leg

After calibrating the measurement unit according to the guidelines of the manufacturer and lying 10–15 min supine to allow body fluids to stabilize, a sinusoidal current of 200 μA was applied and impedance was recorded at 5, 50, 100, and 200 kHz. Finally, the impedance index was calculated as length (l) in cm squared, divided by impedance (l^2/Z). To derive a value representing the whole body based on segmental measurements (BIA$_{Segmental}$) impedance indices for the arm, trunk and leg were summed.

2.2.4 Total body water

Based on research by several investigators (Koulmann *et al.*, 2000; O'Brien *et al.*, 1999), and following recommendations by Schoeller (2000) total body water was calculated from differences in body weight pre- and post-hypohydration. It was assumed that this difference primarily reflects a loss of total body water as only minor losses of carbon ($\approx$50 g) were expected. However, in order to correct changes in nude body mass for metabolic losses, the latter were calculated according to Mitchell *et al.* (1972).

2.2.5 Biochemical assays

Two blood samples of 5 ml and 1.5 ml were drawn from the left antecubital vein using a Beckton Dickonson blood collection set (Vacutainer Systems, Beckton

Dickinson, France). After discarding the first 1ml, a sample of 10 ml was collected in a heparinized plastic syringe without stasis. Subsequently, the sample was split by transferring it into two types of tubes, one of them containing lithium-heparin (BD Vacutainer® K2E containing EDTA; and BD Vacutainer® SST). The latter one was centrifugated at 62360 g for 10 min (GT 422, Juan). Resulting plasma was analyzed for electrolytes (sodium, chloride, potassium) by ion-sensitive electrodes (EML™ Electrolyte Laboratory, RADIOMETER, Copenhagen, Denmark). The sample of the second tube was then analyzed for hematocrit (Hct) and hemoglobin concentration ([Hb]) using the Coulter principle (Coulter Counter® MD8, Coulter Electronics GmbH, Germany). Hct was corrected for trapped plasma (Hct × 0,96) but not for whole-body Hct.

2.2.6 Blood Pressure and resting heart rate

Blood pressure and resting heart rate was measured using the oscillometric Riva-Rocci method with an upper arm cuff (Bosotron 2, Bosch & Sohn, Jungingen, Germany).

2.2.7 Skin surface temperature

Skin surface temperatures of the chest, biceps, anterior thigh, and medial calf were measured by infrared thermometry (Dermatemp 1001, Exergen, MA, USA). Mean skin surface temperature was calculated according to the method of Ramanthan (1964).

2.3 Data Analysis

All statistical analyses were performed using the SPSS software (Version 12.00, SPSS Inc., Illinois, USA). Values were reported as means and standard deviations.

A typical 2-way mixed ANOVA model (subjects were treated as a random effect and trials as a fixed effect) was used to assess the reliability of impedance measurements. Intraclass correlation coefficients (ICC) were calculated from the ratio of between-subject variance and total variance. Standard errors of measurement (SEM) were calculated as $SEM = SD \times (1-ICC)^{0.5}$, where SD is the average standard deviation of impedance measurements. SEM was expressed both in absolute and relative terms, i.e. as coefficient of variation.

Repeated Measures ANOVA was employed to assess between-group (male/female) and within-subject changes over time in the variables of interest as a function of the exercise intervention. Univariate analyses were used to follow-up any significant multivariate effects indicating between group and or within-subject differences. When a significant F-ratio occurred, Scheffé *post-hoc* tests were used to follow-up differences. Correlational analyses were performed to investigate the relationships between changes in TBW and changes in impedance indices. Finally, multiple linear regression analysis was used to examine the potential influence of covariates (Na^+, P^+, and Cl^-, Hct, skin surface temperature, resting heart rate, and blood pressure) on impedance measurements. The null hypothesis was rejected if $P < 0.05$.

3. RESULTS

Table 1.1 and 1.2 show the subjects' general characteristics of the reliability and hypohydration group, respectively. No significant differences for anthropometry and impedance measurements were found between the reliability and hypohydration group ($P > 0.05$). Men were significantly taller, heavier, had a higher impedance index, and a higher $\dot{V}O_{2Max}$ (hypohydration group) than women ($P < 0.001$).

Table 1.1 Descriptive characteristics of reliability group [mean (±SD)]

	Men (n = 10)	Women (n = 10)	Total (n = 20)
Age (years)	25.0 (3.7)	23.2 (3.9)	24.1 (3.8)
Height (m)	179.0 (3.9)	167.9 (7.0)	173.5 (7.9) *
Body mass (kg)	75.3 (9.4)	60.8 (5.5)	68.05 (10.6) *
BMI	23.5 (2.6)	21.5 (1.1)	22.5 (2.2) *
WB H^2/Z (cm^2.Ω^{-1})	72.2 (7.2)	49.9 (6.0)	61.05 (13.1) *

BMI, body mass index; WB, whole body impedance index at 50 kHz. * Significant difference between men and women ($P < 0.05$).

Table 1.2 Descriptive characteristics of hypohydration group [mean (±SD)]

	Men (n = 23)	Women (n = 25)	Total (n = 48)
Age (years)	25.1 (4.6)	22.4 (3.7)	23.8 (4.3) *
Height (m)	180.4 (4.7)	168.4 (6.7)	174.4 (8.3) *
Body mass (kg)	74.8 (7.8)	62.6 (7.3)	68.7 (9.7) *
BMI	23.0 (2.1)	22.1 (2.1)	22.6 (2.1)
WB H^2/Z (cm^2.Ω^{-1})	66.8 (8.5)	50.0 (6.5)	58.4 (11.3) *
VO_{2max} (ml.kg^{-1}.min^{-1})	55.6 (4.7)	47.5 (4.1)	51.6 (6.0) *

BMI, body mass index; VO_{2max}, maximal oxygen uptake; WB, whole body impedance index at 50 kHz. * Significant difference between men and women ($P < 0.05$).

Results from the reliability study demonstrated a very high degree of reproducibility for three successive impedance measurements separated by 60 min over a three hour period. As documented in Table 1.3, the ICC exceeded 0.942 for all measurements and the SEM ranged between 0.3 Ω and 3.6 Ω.

Table 1.3 Reliability impedance measurements [ICC, SEM. CV (SEM%)]

	5 kHz	50 kHz	100 kHz	200 kHz
Whole-body	0.998	0.999	0.999	0.999
	3.6 Ω	2.5 Ω	2.4 Ω	2.3 Ω
	0.6%	0.5%	0.5%	0.5%
Arm	0.997	0.999	0.999	0.999
	2.4 Ω	1.4 Ω	1.3 Ω	1.3 Ω
	0.8%	0.5%	0.6%	0.6%
Trunk	0.973	0.942	0.986	0.981
	0.4 Ω	0.6 Ω	0.3 Ω	0.3 Ω
	1.5%	2.6%	1.4%	1.4%
Leg	0.997	0.998	0.998	0.997
	2.2 Ω	1.6 Ω	1.6 Ω	1.8 Ω
	0.8%	0.7%	0.7%	0.9%

ICC, Intraclass correlation coefficient; SEM, standard error of the measurement.

Body weight corrected for carbon loss (TBW) was significantly reduced from Baseline to Post 0 h and from Post 0 h to Post 1 h in both men and women, respectively ($P < 0.001$). The interaction between time and gender turned out to be significant from Baseline to Post 0 h ($P < 0.05$). However, since the interaction was due to only marginally higher fluid loss in the men and no significant gender × time interaction was found for any of the other outcome measures ($P > 0.05$) the following results are reported for the total group only. Whole body impedance significantly decreased from Baseline to Post 0 h ($P < 0.05$) and significantly increased ($P < 0.05$) from Post 0 h to Post 1 h at 5, 50, 100 and 200 kHz (Figure 1.7).

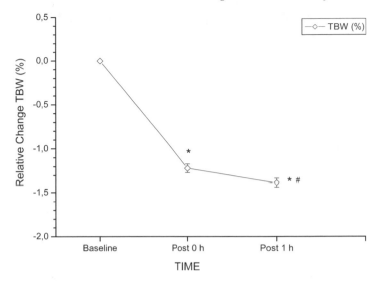

Figure 1.6 Mean (±SEM) percentage change in TBW immediately after and 1 hour after exercise. *Significantly different from Baseline (P < 0.05). # Significantly different from Post 0 (P < 0.05).

Though the time × frequency interaction term turned out to be significant (P < 0.05), the visual inspection of the data indicated an equivalent pattern for all frequencies when impedances were plotted vs. time (Figure 1.7).

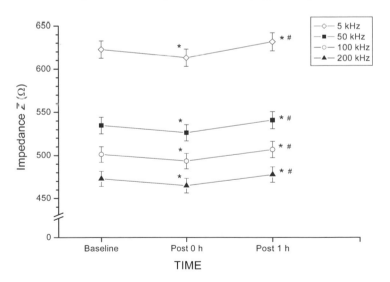

Figure 1.7 Mean (± SEM) change in whole-body impedance immediately after and 1 hour after exercise. * Significantly different from Baseline (P < 0.05). # Significantly different from Post 0 (P < 0.05).

A similar pattern was observed for segmental impedances (Figure 1.8–1.10).

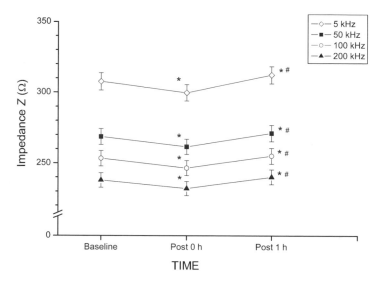

Figure 1.8 Mean (± SEM) change in arm impedance immediately after and 1 hour after exercise. * Significantly different from Baseline ($P < 0.05$). # Significantly different from Post 0 ($P < 0.05$).

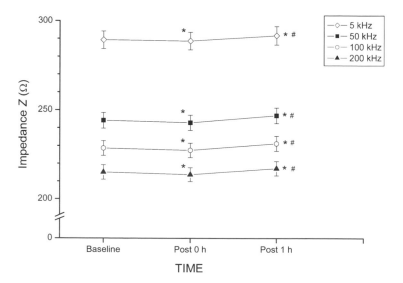

Figure 1.9 Mean (± SEM) change in leg impedance immediately after and 1 hour after exercise. * Significantly different from Baseline ($P < 0.05$). # Significantly different from Post 0 ($P < 0.05$).

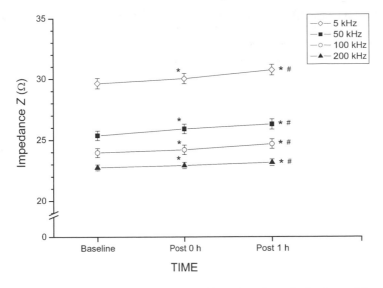

Figure 1.10 Mean (± SEM) change in trunk impedance immediately after and 1 hour after exercise. * Significantly different from Baseline (*P* < 0.05). # Significantly different from Post 0 (*P* < 0.05).

Arm and leg impedance first significantly decreased both from Baseline to Post 0 h (*P* < 0.05), and then increased from Post 0 h to Post 1 h (*P* < 0.05). Trunk impedance showed a significant trend to increase from both Baseline to Post 0 h and Post 0 h to Post 1 h (*P* < 0.05), which, however, was practically negligible.

The lack of association between TBW and BIA was confirmed by correlational analyses. As documented in Table 1.4 neither changes in whole body ($\Delta BIA_{Whole-Body}$) nor segmental ($\Delta BIA_{Segmental}$) impedance indices were related to changes in TBW (ΔTBW).

Table 1.4 Relationships between ΔTBW and $\Delta Impedance$ Indices [r]

	ΔTBW	5 kHz	50 kHz	100 kHz	200 kHz
ΔWB	Baseline – Post 0 h	0.08	0.06	0.008	0.09
	Post 0 h – Post 1 h	0.25	0.27	0.27	0.25
	Baseline – Post 1 h	0.06	0.03	0.05	0.04
ΔSEG	Baseline – Post 0 h	0.20	0.12	0.03	0.02
	Post 0 h – Post 1 h	0.18	0.22	0.22	0.02
	Baseline – Post 1 h	0.28	0.20	0.22	0.09

r, Pearson correlation coefficient; ΔTBW, Change of total body water; ΔWB, Change of whole impedance indices; ΔSEG, Change of segmental impedance indices. None of the correlations was significant (*P* > 0.05).

Except for potassium, all measured variables significantly changed from baseline to Post 0 h after exercise ($P < 0.05$). Subsequently, only resting heart rate significantly changed from Post 0 h to Post 1 h ($P < 0.05$). All other variables remained unaltered between Post 0 h and Post 1 h (Table 1.5).

Table 1.5 Physiological responses to treadmill exercise [mean (±SD)]

	Baseline	Post 0 h	Post 1 h
Skin surface temperature (°C)	31.6 (0.7)	32.2 (0.8) *	32.2 (0.6) *
Resting HR (min⁻¹)	62 (12)	74 (11) *	66 (10) *[#]
SBP (mmHg)	124 (10)	115 (6) *	115 (8) *
DBP (mmHg)	73 (9)	69 (10) *	68 (8) *
Hct (%)	38 (3.1)	39 (3.6) *	39 (3.4) *
Potassium (mmol l⁻¹)	4.1 (0.3)	4.2 (0.3)	4.2 (0.3)
Sodium (mmol l⁻¹)	139.5 (1.5)	141.3 (1.6) *	141.1 (1.5) *
Chloride (mmol l⁻¹)	100.6 (1.2)	101.8 (1.7) *	101.9 (1.3) *

HR, heart rate; RR_{SYS}, systolic blood pressure; RR_{DIA}, diastolic blood pressure; Hct, hematocrit. * Significantly different from baseline ($P < 0.05$). # Significantly different from Post 0 h ($P < 0.05$).

When multiple regression analysis was used to explain impedance changes, none of the covariates (Na^+, P^+, and Cl^-, Hct, skin surface temperature, resting heart rate, and blood pressure) turned out to significantly account for any variance in impedance ($P > 0.05$).

4. DISCUSSION

The aim of the present study was to investigate the usefulness of whole-body and segmental bioelectrical impedance analysis to monitor changes in TBW as a result of exercised-induced hypohydration in a large sample of healthy men and women. Prior to the hypohydration trial a reliability study was performed to assess the within-day variations of BIA measurements with regard to the particular time frame of the hypohydration protocol.

Results from the reliability study showed excellent agreement between three successive measurements separated by 1 hour (Table 1.3) and confirm the minimal within-day variation of 0.6% to 2% (CV) found in previous studies (Kushner and Schoeller, 1986; Lukaski *et al.*, 1985; Segal *et al.*, 1985). Similarly, ICCs ranged between 0.942 and 0.999 in the present study and were well in accordance with results documented by other investigators when short periods of within-day variation were studied (Fornetti *et al.*, 1999; Vehrs *et al.*, 1998).

Assuming that electrical resistivity remains approximately stable during hypohydration, an increase in impedance would be expected after exercise as hypohydration reduces the amount of fluid transmitting the current. In the present study, however, impedance first decreased following exercise and then rose back to baseline levels after an hour of rest. This pattern was observed irrespective of gender, at low and high frequencies and for both whole-body and segmental measurements (Figure 1.6–1.9). Our hypothesis, that a regional redistribution of fluids following exercise masked a true increase in impedance, could therefore not be confirmed. Only trunk impedance showed a continuous increase from baseline to immediately after and 1 h after exercise (Figure 1.10). Though significant, this increase was marginal and clearly within the measurement error of the BIA instrument, and thus should not be considered as a real change. In accordance with these findings no relationships between ΔTBW and $\Delta(l^2/Z)$ could be established (Table 1.4).

Secondly, it was postulated that impedance measurements at higher frequencies would provide a better estimate of TBW after exercise as the applied current does not penetrate all cells at low frequencies (see introduction). Several researchers have argued that the prediction of TBW at a frequency of 50 kHz is based on the high intercorrelation between ECW and TBW (e.g., Matthie *et al.*, 1998) and, consequently, the predictive accuracy of BIA at 50 kHz is limited to conditions in which the relationship between ECW and TBW is unaltered. Schoeller (2000) found that when infusion of Lactated Ringer's solution was used to increase ECW with little change in intracellular water (ICW), TBW was overestimated by resistance at a frequency of 50 kHz. In contrast, resistance at infinite frequency, extrapolated from a Cole-Cole model, could be used to monitor TBW with excellent accuracy. Similarly, any alterations of the equilibrium between fluid compartments resulting from exercise might bias TBW-measurements when BIA measurements at relatively low frequencies (i.e. 50 kHz) are employed. Most investigators reported a pronounced decrease of ECW compared to ICW after bouts of intermediate exercise between 30 min and 2 h (for review see Senay, 1998).

However, it should be noted that the distribution of losses among the fluid compartments can also vary with type of hypohydration, type of exercise, aerobic fitness, acclimatization, initial hydration levels, age, and time point of measurement (for review see Senay, 1998). In spite of the assumed larger loss of ECW in the present study, the advantage of using higher over lower frequencies could not be confirmed. Whether frequencies >200 kHz would have shown an improvement is questionable, as higher frequencies have not always substantially increased the accuracy of BIA-estimated TBW (e.g., Hannan *et al.*, 1995; van Marken Lichtenbelt *et al.*, 1994). However, in this regard it is worth noting that other investigators have argued that the use of any single frequency is problematic because this would require the proportion of ECW and ICW measured at a single frequency to be fixed and the resistivity of TBW to be constant (De Lorenzo *et al.*, 1997a; Matthie *et al.*, 1998). Since the characteristic frequency can vary considerably not only between but also within individuals (De Lorenzo *et al.*, 1997a) the extent to which the applied current is conducted by the ECW and ICW also changes. As a result different physical quantities, i.e. different proportions of TBW are measured in each individual when a single, fixed predetermined frequency is used for impedance measurements.

Cornish and associates (1993) therefore suggested that the use of impedance at the characteristic frequency should improve the estimation of TBW as the ratio of the currents that pass through the extra- and intracellular fluids is independent of capacitance at this frequency and therefore the same physical quantities are measured in each subject (Cornish *et al.*, 1993). However, according to De Lorenzo *et al.* (1997a) this does not alleviate the problem that using any single frequency is based on the assumption that the resistivity of TBW is constant. This assumption can be violated by a simple change in the ratio of ECW and ICW. Given the fact that the resistivity of intracellular fluid is approximately 4 to 5 times higher than the resistivity of extracellular fluid any alterations of the ECW–ICW ratio will affect the overall resistivity term for TBW even if the resistivities of intra- and extracellular fluid volumes remain unaltered. De Lorenzo *et al.* (1997a) therefore suggested determining the resistances of intra- and extracellular fluids separately, using them for the determination of ECW and ICW, respectively, and finally obtaining a measure of TBW by adding the two compartments. Though this approach seems appealing it is prone to a number of errors as eloquently outlined by Ward *et al.* (1998). However, even ignoring that the true advantage of mixture theory in modeling fluid volumes still needs to be established (Ellis *et al.*, 1999; Gudivaka *et al.*, 1999; Martinoli *et al.*, 2003; Schoeller, 2000) the observed post-exercise decrease in impedance in the present study would have also been mirrored by any other BIA approach and thus similarly affected its success in predicting TBW.

In contrast to our findings, several researchers have documented advantages of segmental over whole-body BIA to monitor fluid shifts. For instance, Patterson *et al.* (1988) found that a segmental approach improved the detection of fluid shifts during dialysis considerably. These results were further substantiated by Zhu *et al.* (2000). Moreover, Bracco *et al.* (2000) showed that segmental BIA was able to detect and localize perioperative fluid accumulation. The reason for the failure of segmental BIA at 200 kHz to accurately monitor changes in TBW in the present study must therefore be related to the physiological perturbations caused by aerobic exercise.

According to the authors' knowledge, 16 studies have been conducted on the effect of physical activity on impedance which are summarized in Table 1.6. Additionally, Cornish *et al.* (2001) investigated the effects of a 3-week exercise program on fluid shifts measured by BIA in rats. Furthermore, Koulmann *et al.* (2000) studied the feasibility of monitoring exercise-induced fluid losses by BIA and found a 50% underestimation of BIA-predicted TBW. However, as raw values of impedance were not reported, it remains speculative whether impedance truly increased or other predictors underlying the algorithm used to estimate TBW from BIA (e.g., body mass) confounded the results. Similarly, Goss *et al.* (2003) who performed leg-to-leg BIA before and after a maximal graded exercise test in children found that percentage body fat slightly decreased after exercise. Again, however, whether this observation can be attributed to a decrease in impedance cannot be determined for impedance data were not reported. The same problem pertains to the studies by Kleiner (1992) and Jurimae *et al.* (2000).

Table 1.6 Summary of studies of the effects of physical exercise on BIA

Study	n	Frequency (kHz)	Z, R
Present study	46	5, 50, 100, 200	↓
Pialoux *et al.* (2004)	12	100	↓
Demura *et al.* (2002)	30	50	↓
O'Brien *et al.* (1999)	9	BIS	↑
Fellmann *et al.* (1999)	9	5, 100	↓
Saunders *et al.* (1998)	15	50	↑
Asselin *et al.* (1998)	3	5, 50, 100	↓
Monnier *et al.* (1997)	11	1, 10, 50, 100	↓
Collodel *et al.* (1997)	10	50	↔
Liang and Norris (1993)	30	50	↓[*]
Thompson *et al.* (1991)	10	50	↓
Garby *et al.* (1990)	5	50	↔
Deurenberg *et al.* (1988)	9	50	↓
Abu Khaled *et al.* (1988)	8	50	↓
Stump *et al.* (1988)	14	50	↔
Schell and Gross (1987)	12	50	↓[*]

Z, impedance; R, Resistance; BIS, Bioimpedance Spectroscopy using Cole-Cole modelling to extrapolate resistance at 0 and infinite frequency. *In some subjects an increase in impedance was observed.

Three investigations, Collodel *et al.* (1997), Garby *et al.* (1990), and Stump *et al.* (1988), found no change in impedance after physical activity which can probably be attributed to the short exercise protocols employed in these studies and/or a lack of power to detect significant differences[1]. The majority of studies observed a decrease in impedance following exercise and are in accordance with the results of the present study. It is assumed that resistivity (inverse of electrical conductivity) changes as a result of hypertonic hypohydration.

Interestingly, in contrast to the decrease in impedance found in the majority of studies, Saunders *et al.* (1998) and O'Brien *et al.* (1999) noted an increase in impedance following exercise. Moreover, O'Brien *et al.* (1999) found a moderate to strong relationship ($r = 0.77$; $P < 0.05$) between the BIA-estimated changes in TBW and changes in TBW calculated by the difference of TBW determined by isotope dilution and body mass loss. They concluded that BIA (BIS) is valid for young, fit male subjects in both euhydrated and hypohydrated states and that BIA

---

[1] Furthermore, a continuous increase in impedance was observed during the recovery phase in the study by Stump *et al.* (1988), supporting the general trend found in the other studies.

(BIS) is able to monitor changes following hypertonic hypohydration[2]. Similarly, Fellmann *et al.* (1999) and Pialoux *et al.* (2004) also confirmed that BIA is sufficiently accurate to monitor TBW. The reason for the discrepancy between these promising findings and the results from the majority of studies, including the present investigation, can be explained by different research protocols. Fellmann *et al.* (1999), Pialoux *et al.* (2004) and O'Brien *et al.* (1999) performed the BIA measurement not immediately after exercise, but after an extensive period of rest (varying between an overnight rest and several days of recovery). Furthermore, in the studies by Fellmann *et al.* (1999) and Pialoux *et al.* (2004) subjects were not only allowed to consume water and food *ad libitum* during the exercise period but also returned to their individual diets during the recovery period[3]. Thus, any physiological changes caused by physical activity that might confound the measurement might have been compensated by sufficient rest and/or appropriate fluid and food intake. However, the findings by O'Brien *et al.* (1999) deserve closer attention as the higher accuracy to monitor fluid shifts by BIA (BIS) during hypertonic hypohydration compared to isotonic hypohydration is less clear. O'Brien *et al.* (1999) also investigated the effects of isotonic fluid loss on BIA and found that in contrast to hypertonic hypohydration ($r = 0.77$; $P < 0.05$) changes in BIA (BIS) determined TBW were not significantly related to changes in TBW measured by isotope dilution ($r = 0.07$). It was argued that increased tonicity following hypertonic fluid loss might improve the validity of BIA (BIS) for TBW determination as "increased plasma tonicity during HH [hypertonic hypohydration] would increase electrical conductivity, resulting in a corresponding reduction in TBW$_{BIS}$. The blunted increase in plasma tonicity after IH [isotonic hypohydration] was expected to reduce the ability of BIS to detect the change in TBW". This conclusion is, however, questionable. Any significant change in electrical conductivity will jeopardize the technique underlying BIA because resistivity is assumed to be constant in BIA and therefore the proportional removal of fluid and electrolytes is an essential condition to yield accurate estimates of TBW. In other words, BIA should only provide a valid measurement of TBW when plasma osmolality, electrolyte concentrations and hematocrit are unchanged (Pialoux *et al.*, 2004; Rees *et al.*, 1999; Thomas *et al.*, 1999). The failure of estimating TBW by BIA during isotonic hypohydration compared to hypertonic hypohydration reported by O'Brien *et al.* (1999) is therefore puzzling. Isotonic hypohydration would be expected to reflect BIA-estimated fluid volume changes more accurately than during hypertonic hypohydration since ionic concentrations are relatively stable and any physiological perturbations related to physical exercise are irrelevant during isotonic hypohydration. This is in accordance with findings from several investigators who found BIA to accurately monitor ECW and TBW after diuretic treatment (e.g., Gudivaka *et al.*, 1999; Schoeller, 2000).

---

[2] Though O'Brien *et al.* (1999) showed that TBW measured by isotope dilution accounted for 70% of the variance in BIA (BIS) after hypertonic hypohydration and the changes in TBW determined by both methods were also strongly correlated ($r = 0.77$), significant differences in the amounts of TBW loss (1.4 L) between the two techniques point out the weakness of the technique to monitor acute fluid changes.

[3] In the study by O'Brien *et al.* (1999) fluid replacement during exercise was only limited to small amounts of water, but not totally restricted.

Given that furosemide primarily induces a loss from ECW, the concomitant change of the ECW–ICW ratio could have changed the resistivity of TBW for reasons discussed above.

However, though not explicitly specified, it is assumed that O'Brien and colleagues determined ECW and ICW independently using mixture theory because this approach is integrated in the Xitron software which was reported to be used for the evaluation of impedance data. The discrepancy between the results observed by hypertonic and isotonic hypohydration might therefore simply underline the general limitations of present BIA principles and models to accurately monitor relatively small fluid volume changes. However, such a generalization does not appear prudent or justified as a number of other researchers have verified the potential of BIA to monitor fluid changes accurately under various clinical conditions (Bracco *et al.*, 2000; Patterson *et al.*, 1988; Pialoux *et al.*, 2004, Zhu *et al.*, 2000). Again, it should be stressed that not a change in resistivity, but its constancy or respective correction (Scharfetter *et al.*, 1997), under conditions where resistivity changes, is a prerequisite underlying any BIA-model. Accordingly, as already mentioned above, it is assumed that the decrease in impedance found in the present study was related to an increase in resistivity concomitant with aerobic exercise.

Resistivity is directly proportional to ion concentration (and their limiting equivalent conductivity). Additionally, resistivity is influenced by the solution viscosity, the amount of non-conducting material in the solution, and temperature. Therefore, an increase in electrolyte concentrations, an increase in skin and body temperature and a change in hematocrit, glucose, proteins and lactate following a hypertonic hypohydration may potentially affect BIA.

Skin temperature slightly, but significantly, increased (0.6 °C) from baseline to immediately after exercise ($P < 0.05$). Caton *et al.* (1988) reported impedance to be decreased when skin temperature was increased. It was hypothesized that this temperature-dependent effect was due to alterations in cutaneous blood flow and/or a compartmental distribution of body water. Though blood flow might not have directly influenced impedance a concomitant vasodilation could have caused an increase of the vascular volume and thus a drop in impedance. On the other hand, Gudivaka *et al.* (1996) also found an inverse relationship between skin temperature and whole-body impedance (temperature coefficient of -1%/°C), but excluded the effects of increased skin perfusion, vasodilation or an expansion of ECW. In contrast, they hypothesized that the observed decrease in impedance was predominantly the result of an altered skin-electrode impedance. These results, however, directly contradict the findings reported by Buono *et al.* (2004) who suggested that electrode-skin interface temperature does not affect impedance measurements. This is in agreement with other investigators who concluded that the effects of skin temperature on impedance are insignificant (Cornish *et al.*, 1998) or at least minimal (Liang and Norris, 1993) and do not explain the large prediction error of fluid loss following physical activity (Koulmann *et al.*, 2000). Interestingly, Cornish *et al.* (1998) found that increased skin temperature, in fact, has an effect on impedance, however, only when a bipolar electrode arrangement is employed and this influence can be offset by using a tetrapolar electrode configuration which is negligibly affected by skin impedance (less then 2% when skin temperature changed by 15 °C). Similarly, the same authors demonstrated that though changes in skin hydration can considerably influence skin resistance, the resistance of the underlying body tissue remains unaltered when current and voltage are introduced and sensed via two separate electrode pairs. Furthermore, sweat rate has been shown to decrease by 50% after 5 min of post-exercise rest and to return to baseline levels after approximately 15 min (Journeay *et al.*, 2004). Considering that in the present study subjects were first dried and changed clothes after exercise and impedance measurements were only taken after 10 min of lying supine and electrode sites were cleaned with a 70% alcohol solution prior to measurements, it is unlikely that impedance was substantially affected by either sweating or an elevated skin temperature. It should be noted that core temperature, which was not measured in the present study, remains as a potential source for the observed decrease in impedance. Though the increase in core temperature would be expected to be relatively small (about 1 °C after one hour of exercise) and also to decrease sharply after termination of exercise, internal limb temperature could have increased to a larger extent and only gradually declined during recovery (Kenny *et al.*, 2003). It can therefore be speculated that internal limb temperature could have caused a decrease in impedance despite minor alterations in skin and core temperature.

Resting heart rate was also significantly elevated whereas blood pressure was significantly decreased at Post 0 h ($P < 0.001$). However, multiple regression analysis showed that none of these parameters explained the post-exercise decrease in impedance. In accordance with this finding, Liang and Norris (1993) concluded that BIA measurements were only minimally affected by substantial changes in regional and mean skin blood flow subsequent to physical activity.

Hematocrit is inversely related to conductivity (Jaspard *et al.*, 2003). In the present study hematocrit was significantly increased after exercise ($P < 0.05$). Irrespective of the fact that the increase of hematocrit was trivial ($\approx 1\%$), a higher hematocrit would have increased and not decreased impedance.

Thus, only the changes in blood chemistry remain as potential covariates to explain the findings in the present study. Due to the hypotonic characteristic of sweat an increase in ion concentrations was expected. Sodium, chloride, potassium were chosen because they are the major extracellular and intracellular electrolytes, respectively and they represent the dominant electrolytes in sweat (Senay, 1998). Sodium, chloride and potassium all increased from baseline to immediately after exercise. However, the change in potassium was not significant. Moreover, neither sodium nor chloride changes were significantly related to changes in impedance. In contrast, Asselin *et al.* (1998) noted that potassium increased subsequent to exercise and explained 37% of the variation in impedance. However, it should also be noted that they used a combination of exercise and heat stress to induce hypohydration and, consequently, skin temperature increased by 6 °C. Thus, both increased skin temperature and its subsequent considerable increase in skin blood flow could have explained the post-exercise decrease in impedance as well. Finally, though significant the increase in sodium, chloride and calcium were so small that it is questionable whether they truly could have affected impedance. Rees *et al.* (1998) reported that a significant influence of ion concentrations on impedance was only observed after a 12% sodium increase of total blood sodium. In contrast, sodium increased by 1.2% in the present study.

In summary, except for potassium all potential covariates demonstrated a significant change after exercise (Table 1.5). However, except resting heart rate which was increased immediately after exercise (Post 0 h) and then returned to baseline levels an hour later (Post 1 h) all other variables showed either a decrease or increase at Post 0 h which remained unaltered during recovery. This observation (ignoring resting heart rate) demonstrates the difficulty of relating any of the variables to the observed changes in impedance (Figure 1.7–1.10) for they do not seem to account for any variation in impedance at Post 1 h.

As indicated at the beginning of the discussion, resistivity of TBW can also change even if ion concentrations or any other physiological influential factors do not change. As outlined above, any alterations of the ECW–ICW ratio can alter the resistivity of TBW. Thus, a pronounced decrease of fluid volume from the extracellular compared to the intracellular compartment could have decreased the resistivity of TBW and caused the observed drop in impedance immediately after exercise. Furthermore, the imbalance between ECW and ICW might be slightly alleviated during recovery as the impact of the Starling forces (i.e. particularly the decrease of the hydrostatic pressure) is reduced and the effect of osmotically active particles in the muscles is also lessened (and at the same time allows the protein oncotic pressure to exert its influence on the fluid distribution between the intra- and extravascular volumes). In turn, resistivity of TBW might have increased again which could explain the rise of impedance back to baseline levels. Though the described physiological pertubations would be in accordance with the elevated heart rate after exercise and its decline to baseline levels at Post 1 h (heart rate increases to compensate the lower plasma volume resulting from a change in the Starling forces), it should be noted that this scenario is highly speculative. Firstly, it has been argued that the post-exercise recovery of plasma is simply an exchange of fluids between the extra- and intravascular compartments and thus do not affect ICW (Nose *et al.*, 1988). It could also be argued that the increase in plasma proteins immediately after exercise increases fluid viscosity and thus decreases resistivity by reducing ion mobility (Schoeller, 2000). Moreover, we found that the ratio of impedance at low (5 kHz) and high frequencies (200 kHz), which can be used as a measure of compartmental water shifts (De Lorenzo *et al.*, 1997b), did not significantly change ($P > 0.05$) at any time point. Accordingly, it could be concluded that the ECW–ICW was stable during the intervention. Though this is not impossible it appears to contradict the assumption that the lost fluid was primarily drawn form the extracellular fluid and thus rather underlines the insensitivity of the ratio of impedance at 5 kHz and 200 kHz to small water shifts between the subcompartments. It cannot also be excluded that the constant ratio at low and high frequencies was the result from a number of counteracting forces (e.g., a relatively extensive loss of fluid volume from one compartment that was offset by a concomitant increase in resistivity of that fluid compartment).

Finally, it is unlikely that the measurement was confounded by the protocol, i.e. by ambient temperature, electrode placement, limb position and posture. Irrespective of the high reliability of BIA measurements found in this study, measurements were performed in a thermally controlled environment (20–22 °C ambient temperature and 55–60% relative humidity). To increase intra-observer reliability with regard to electrode placement, electrode sites were marked prior to the initial measurement. Limb positions were also fixed by using styrofoam positioning blocks that defined arm and leg abduction, respectively. To account for fluid shifts from standing to supine measurements were only performed after 15 min. Furthermore, segmental BIA has been shown to be unaffected by orthostatic fluid shifts compared to whole-body BIA (Thomas *et al.*, 1998).

The question of what caused post-exercise impedance values to be elevated thus remains unclear. Koulmann *et al.* (2000) suggested that water bound to the glycogen matrix might not be detectable by BIA. Therefore, glycogen breakdown during physical activity and its subsequent release of water could potentially mask the decrease in BIA-estimated TBW. Another possibility would be the effect of breathing on impedance. Gualdi-Russo and Toselli (2002) showed that impedance at 50 kHz can increase during maximum inspiration and expiration by 13 and 21.5 Ω in men and women, respectively, compared to normal breathing. Hence, an elevated pulmonary ventilation after exercise could have potentially influenced the measurements but would neither, however, explain the observed decrease in impedance. It could also be speculated that the violation of the basic assumption that the human body is an isotropic conductor leads to misinterpretations in the estimation of TBW. Though this assumption is reasonable for the limbs where muscle fibres run parallel to the measurement axis, the trunk's anatomy is considerably more complex. Apart from the number of different organs muscle fibres also run perpendicular to the measurement axis. Since transverse conductivity of muscles is significantly higher than their longitudinal conductivity (Aaron *et al.*, 1997) the underlying biophysical model of BIA might not reflect small changes in trunk volume, even when a segmental approach is employed. Additionally, the fluid shift might have been simply too small in comparison to the volume of the trunk (Thomas *et al.*, 1999). Furthermore, in spite of any significant change in the ratio of impedance at 5 kHz and 200 kHz it is possible that an imbalance of ECW and ICW was prevalent in the corium of the skin and caused a "local" decrease in the overall resisitivity of the two subcompartments. Lastly, it seems noteworthy that the intraclass coefficient as well as the coefficient of variation from the reliability study have to be interpreted with care since individual deviations between two measurements were much higher than the average statistics reported in Table 1.3. Though most of the individual differences were quite small ($<5 \Omega$) at 200 kHz some differences were twice as great. Interestingly, the trend showed a clear decrease in impedance which is difficult to interpret. It could reflect a combination of evaporation via expired air and the skin ($\approx$20–35 ml.h^{-1}) and renal water losses ($\approx$50–100 ml.h^{-1}), which, however, cannot be verified.

5. CONCLUSION

In conclusion, it is unlikely that there is a single primary factor that can be associated with observed decrease in impedance after exercise. Assuming that temperature might have played a role in the present study, it is safe to assume that other factors also account for the influence on impedance. Thus in spite of the failure of multiple regression analysis to identify any significant potential covariates a number of factors, such as skin hydration, skin surface, internal limb or body temperature, skin blood flow, an increased ion concentrations concomitant with hypertonic hypohydration, and a regional fluid redistribution might have contributed to the impact on impedance data. As the sample size was fairly large in the present study, it is suggested that the insignificance of potential covariates affecting BIA is not due to a lack of power, but rather to the fact that the extent of each influential factor might be different between individuals. This was also pointed out by Liang and Norris (1993) who reported that in spite of an average decrease in impedance following exercise, impedance increased in 13% of their subjects. Similarly, Schell and Gross (1987) found impedance on average to decrease after 2 h of volleyball playing. However, they also observed that in 3 out of 12 subjects impedance increased. Moreover, Kanai *et al.* (1987) also found both a decrease and increase in subjects after several types of exercise. As the effect of exercise on impedance was dependent on the subjects' aerobic fitness (increase in trained, decrease in untrained subjects) they suggested impedance to be an indicator of post-exercise recovery and thus athletic ability. The assumption of differential effects of exercise on impedance measurements with regard to aerobic fitness seems plausible (e.g. sweat rate, sweat composition, temperature regulation, increased TBW, blood volume and ECW) and we therefore also performed a case-wise inspection of impedance data with regard to aerobic capacity. The findings by Kanai *et al.* (1987) could, however, not be confirmed as no pattern with regard to athletic status was apparent. Whether this was due to the homogeneous nature of the sample and the fact that all subjects were engaged at least in some physical activity programmes cannot be verified. The case-wise inspection, however, clearly revealed a substantial variation in post-exercise impedance data. It is therefore concluded that BIA is sensitive to a number of physiological changes occurring after acute bouts of exercise and that the impact of these changes might vary interindividually and therefore it is hardly possible to account for these influences when impedance measurements are performed.

In summary, the aim of the present study was to investigate the usefulness of whole-body and segmental BIA at a low and high frequencies to monitor a hypertonic hypohydration induced by physical activity. It was found that instead of the expected increase in impedance following exercise, impedance first decreased from baseline to immediately after exercise and returned to approximately baseline values after an hour of recovery. Furthermore, no advantage could be observed for measurements made at higher compared to lower frequencies. Similarly, the segmental approach could not improve the detection of fluid loss. None of the potential covariates known to affect BIA turned out to be significant. It was concluded that the drawbacks involved in conventional whole-body BIA are not off-set by a segmental approach during exercise-induced hypohydration and that BIA is sensitive to a number of factors and their impact might vary inter-individually.

REFERENCES

Aaron, R., Huang, M. and Shiffman, C.A., 1997, Anisotropy of human muscle via non-invasive impedance measurements. *Physics in Medicine and Biology*, **42**, pp. 1245–1262.

Abu Khaled, M., McCutcheon, M.J., Reddy, S., Pearman, P.L., Hunter, G.R. and Weinsier, R.L., 1988, Electrical impedance in assessing human body composition: the BIA method. *American Journal of Clinical Nutrition*, **47**, pp. 789–792.

Asselin, M.C., Kriemler, S., Chettle, D.R., Webber, C.E., Bar-Or, O. and McNeill, F.E., 1998, Hydration status assessed by multi-frequency bioimpedance analysis. *Applied Radiation and Isotopes*, **49**, pp. 495–497.

Baumgartner, R.N., Chumlea, W.C. and Roche, A.F., 1989, Estimation of body composition from bioelectric impedance of body segments. *American Journal of Clinical Nutrition*, **50**, pp. 221–226.

Bracco, D., Berger, M., Revelly, J.P., Schutz, Y., Frascarolo, P. and Chiolero, R., 2000, Segmental bioelectrical impedance analysis to assess perioperative fluid changes. *Critical Care Medicine*, **28**, pp. 2390–2396.

Buono, M.J., Burke, S., Endemann, S., Graham, H., Gressard, C., Griswold, L. and Michalewicz, B., 2004, The effect of ambient air temperature on whole-body bioelectrical impedance. *Physiological Measurement*, **25**, pp. 119–123.

Caton, J.R., Mole, P.A., Adams, W.C. and Heustis, D.S., 1988, Body composition analysis by bioelectrical impedance: effect of skin temperature. Medicine and Science in Sports and Exercise, **20**, pp. 489–491.

Chamney, P.W., Kramer, M., Rode, C., Kleinekofort, W. and Wizemann, V. 2002, A new technique for establishing dry weight in hemodialysis patients via whole body bioimpedance. *Kidney International*, **61**, pp. 2250–2258.

Cian, C., Barraud, P.A., Melin, B. and Raphel, C., 2001, Effects of fluid ingestion on cognitive function after heat stress or exercise-induced hypohydration. *International Journal of Psychophysiology*, **42**, pp. 243–251.

Clarys, J.P. and Marfell Jones, M.J., 1986, Anatomical segmentation in humans and the prediction of segmental masses from intra-segmental anthropometry. *Human Biology; An International Record of Research*, **58**, pp. 771–782.

Collodel, L., Favretto, G., Teodori, T., Caenaro, G., Mordacchini, M., Stritoni, P., Nieri, A. and Piccoli, A. 1997, Use of bioelectrical impedance analysis for monitoring fluid shift during maximal aerobic exercise. *Medicina Dello Sport*, **50**, pp. 197–202.

Cornish, B.H., Thomas, B.J. and Ward, L.C., 1993, Improved prediction of extracellular and total body water using impedance loci generated by multiple frequency bioelectrical impedance analysis. *Physics in Medicine and Biology*, **38**, pp. 337–346.

Cornish, B.H., Ward, L.C., Thomas, B.J., Jebb, S.A. and Elia, M., 1996, Evaluation of multiple frequency bioelectrical impedance and Cole–Cole analysis for the assessment of body water volumes in healthy humans. *European Journal of Clinical Nutrition*, **50**, pp. 159–164.

Cornish, B.H., Thomas, B.J. and Ward, L.C., 1998, Effect of temperature and sweating on bioimpedance measurements. *Applied Radiation and Isotopes*, **49**, pp. 475–476.

Cornish, B.H., Wotton, M.J., Ward, L.C., Thomas, B.J. and Hills, A.P., 2001, Fluid shifts resulting from exercise in rats as detected by bioelectrical impedance. *Medicine and Science in Sports and Exercise*, **33**, pp. 249–254.

De Lorenzo, A., Andreoli, A., Matthie, J. and Withers, P., 1997a, Predicting body cell mass with bioimpedance by using theoretical methods: a technological review. *Journal of Applied Physiology*, **82**, pp. 1542–1258.

De Lorenzo, A., Andreoli, A. and Deurenberg, P., 1997b, Impedance ratio as a measure of water shifts. *Annals of Nutrition & Metabolism*, **41**, pp. 22–28.

Demura, S., Yamaji, S., Goshi, F., Nagasawa, Y., 2002, The influence of transient change of total body water on relative body fats based on three bioelectrical impedance analyses methods. Comparison between before and after exercise with sweat loss, and after drinking. *Journal of Sports Medicine and Physical Fitness*, **42**, pp. 38-44.

Deurenberg, P., Weststrate, J.A., Paymans, I. and van der Kooy, K., 1988, Factors affecting bioelectrical impedance measurements in humans. European Journal of Clinical Nutrition, **42**, pp. 1017-1022.

Ellis, K.J., Bell, S.J., Chertow, G.M., Chumlea, W.C., Knox, T.A., Kotler, D.P., Lukaski, H.C. and Schoeller, D.A., 1999, Bioelectrical impedance methods in clinical research: a follow-up to the NIH Technology Assessment Conference. *Nutrition*, **15**, pp. 874–880.

Fellmann, N., Ritz, P., Ribeyre, J., Beaufrere, B., Delaitre, M. and Coudert, J., 1999, Intracellular hyperhydration induced by a 7-day endurance race. *European Journal of Applied Physiology*, **80**, pp. 353–359.

Fornetti, W.C., Pivarnik, J.M., Foley, J.M. and Fiechtner, J.J., 1999, Reliability and validity of body composition measures in female athletes. *Journal of Applied Physiology*, **87**, pp. 1114–1122.

Garby, L., Lammert, O., Nielsen, E., 1990, Negligible effects of previous moderate physical activity and changes in environmental temperature on whole body electrical impedance. European Journal of Clinical Nutrition, **44**, pp. 545 546.

Goss, F., Robertson, R., Williams, A., Sward, K., Abt, K., Ladewig, M., Timmer, J. and Dixon, C., 2003, A comparison of skinfolds and leg-to-leg bioelectrical impedance for the assessment of body composition in children. *Dynamic Medicine*, **2**, p. 5.

Greenleaf, J.E., 1992, Problem: thirst, drinking behavior, and involuntary hypohydration. *Medicine and Science in Sports and Exercise*, **24**, pp. 645–656.

Gualdi-Russo, E. and Toselli, S., 2002, Influence of various factors on the measurement of multifrequency bioimpedance. *Homo – Journal of Comparative Human Biology*, **53**, pp. 1–16.

Gudivaka, R., Schoeller, D. and Kushner, R.F., 1996, Effect of skin temperature on multifrequency bioelectrical impedance analysis. *Journal of Applied Physiology*, **81**, pp. 838–845.

Gudivaka, R., Schoeller, D.A., Kushner, R.F. and Bolt, M.J., 1999, Single- and multifrequency models for bioelectrical impedance analysis of body water compartments. *Journal of Applied Physiology*, **87**, pp. 1087–1096.

Hannan, W.J., Cowen, S.J., Plester, C.E., Fearon, K.C. and deBeau, A., 1995, Comparison of bio-impedance spectroscopy and multi-frequency bio-impedance analysis for the assessment of extracellular and total body water in surgical patients. *Clinical Science and Molecular Medicine*, **89**, pp. 651–658.

Jaspard, F., Nadi, M. and Rouane, A., 2003, Dielectric properties of blood: an investigation of haematocrit dependence. *Physiological Measurement*, **24**, pp. 137–147.

Journeay, W.S., Reardon, F.D., Martin, C.R. and Kenny, G.P., 2004, Control of cutaneous vascular conductance and sweating during recovery from dynamic exercise in humans. *Journal of Applied Physiology*, **96**, pp. 2207–2212.

92   *Stahn* et al.

Jurimae, J., Jurimae, T. and Pihl, E., 2000, Changes in body fluids during endurance rowing training. *Annals of the New York Academy of Sciences*, **904**, pp. 353–358.

Kanai, H., Haeno, M. and Sakamoto, K., 1987, Electrical measurement of fluid distribution in legs and arms. *Medical Progress through Technology*, **12**, pp. 159–170.

Kenny, G.P., Reardon, F.D., Zaleski, W., Reardon, M.L., Haman, F. and Ducharme, M.B., 2003, Muscle temperature transients before, during, and after exercise measured using an intramuscular multisensor probe. *Journal of Applied Physiology*, **94**, pp. 2350–2357.

Kleiner, D.M., 1992, The validity of bioelectrical impedance analysis in determining exercise induced hypohydration. *Medicine and Science in Sports and Exercise*, **24**, p. S579.

Koulmann, N., Jimenez, C., Regal, D., Bolliet, P., Launay, J.C., Savourey, G. and Melin, B., 2000, Use of bioelectrical impedance analysis to estimate body fluid compartments after acute variations of the body hydration level. *Medicine and Science in Sports and Exercise*, **32**, pp. 857–864.

Kushner, R.F. and Schoeller, D.A., 1986, Estimation of total body water by bioelectrical impedance analysis. *American Journal of Clinical Nutrition*, **44**, pp. 417–424.

Liang, M.T. and Norris, S., 1993, Effects of skin blood flow and temperature on bioelectric impedance after exercise. *Medicine and Science in Sports and Exercise*, **25**, pp. 1231–1239.

Lohman, T.G., Roche, A.F. and Martorell, R. 1988, *Anthropometric standardization reference manual* (Champaign, IL: Human Kinetics Books).

Lollgen, H., Ulmer, H.V. and Crean, P., 1988, Recommendations and standard guidelines for exercise testing. Report of the Task Force Conference on Ergometry, Titisee 1987. *European Heart Journal*, **9 Suppl K**, pp. 3–37.

Lozano, A., Rosell, J. and Pallas-Areny, R., 1995, A multifrequency multichannel electrical impedance data acquisition system for body fluid shift monitoring. *Physiological Measurement*, **16**, pp. 227–237.

Lukaski, H.C., Johnson, P.E., Bolonchuk, W.W. and Lykken, G.I., 1985, Assessment of fat-free mass using bioelectrical impedance measurements of the human body. *American Journal of Clinical Nutrition*, **41**, pp. 810–817.

Martinoli, R., Mohamed, E.I., Maiolo, C., Cianci, R., Denoth, F., Salvadori, S. and Iacopino, L., 2003, Total body water estimation using bioelectrical impedance: a meta-analysis of the data available in the literature. *Acta Diabetologica*, **40**, pp. S203–S206.

Matthie, J., Zarowitz, B., De Lorenzo, A., Andreoli, A., Katzarski, K., Pan, G. and Withers, P., 1998, Analytic assessment of the various bioimpedance methods used to estimate body water. *Journal of Applied Physiology*, **84**, pp. 1801–1816.

Mitchell, J.W., Nadel, E.R. and Stolwijk, J.A.J., 1972, Respiratory weight losses during exercise. *Journal of Applied Physiology*, **32**, pp. 474–476.

Monnier, J.F., Raynaud, E., Brun, J.F. and Orsetti, A., 1997, Influence de la prise alimentaire et de l'exercice physique sur une technique d'impédancemétrie appliqué à la détermination de la composition corporelle. *Science & Sports*, **12**, pp. 256–258.

Nose, H., Mack, G.W., Shi, X.R. and Nadel, E.R., 1988, Shift in body fluid compartments after dehydration in humans. *Journal of Applied Physiology*, **65**, pp. 318–324.

O'Brien, C., Baker-Fulco, C.J., Young, A.J. and Sawka, M.N., 1999, Bioimpedance assessment of hypohydration. *Medicine and Science in Sports and Exercise*, **31**, pp. 1466–1471.

Organ, L.W., Bradham, G.B., Gore, D.T. and Lozier, S.L., 1994, Segmental bioelectrical impedance analysis: theory and application of a new technique. *Journal of Applied Physiology*, **77**, pp. 98–112.

Patterson, R., Ranganathan, C., Engel, R. and Berkseth, R., 1988, Measurement of body fluid volume change using multisite impedance measurements. *Medical & Biological Engineering & Computing*, **26**, pp. 33–37.

Pialoux, V., Mischler, I., Mounier, R., Gachon, P., Ritz, P., Coudert, J. and Fellmann, N., 2004, Effect of equilibrated hydration changes on total body water estimates by bioelectrical impedance analysis. *The British Journal of Nutrition*, **91**, pp. 153–159.

Ramanthan, N.L., 1964, A new weighing system for mean surface temperature of the human body. *Journal of Applied Physiology*, **19**, pp. 531–533.

Rees, A.E., Ward, L.C., Cornish, B.H. and Thomas, B.J., 1999, Sensitivity of multiple frequency bioelectrical impedance analysis to changes in ion status. *Physiological Measurement*, **20**, pp. 349–362.

Saunders, M.J., Blevins, J.E., Broeder, C.E., 1998, Effects of hydration changes on bioelectrical impedance in endurance trained individuals. Medicine and Science in Sports and Exercise, **30**, pp. 885–892.

Sawka, M.N., 1992, Physiological consequences of hypohydration: exercise performance and thermoregulation. *Medicine and Science in Sports and Exercise*, **24**, pp. 657–670.

Sawka, M.N. and Coyle, E.F., 1999, Influence of body water and blood volume on thermoregulation and exercise performance in the heat. *Exercise and Sport Sciences Reviews*, **27**, pp. 167–218.

Scharfetter, H., Wirnsberger, G.H., Holzer, H. and Hutten, H., 1997, Influence of ionic shifts during dialysis on volume estimations with multifrequency impedance analysis. *Medical & Biological Engineering & Computing*, **35**, pp. 96–102.

Schell, B. and Gross, R., 1987, The reliability of bioelectrical impedance measurements in the assessment of body composition in healthy adults. *Nutrition Reports International*, **36**, pp. 449–459.

Schoeller, D.A., 2000, Bioelectrical impedance analysis. What does it measure? *Annals of the New York Academy of Sciences*, **904**, pp. 159–162.

Segal, K.R., Gutin, B., Presta, E., Wang, J. and Van Itallie, T.B., 1985, Estimation of human body composition by electrical impedance methods: a comparative study. *Journal of Applied Physiology*, **58**, pp. 1565–1571.

Senay, L.C. Jr., 1998, Water and Electrolytes During Physical Activity. In *Nutrition in Exercise and Sport*, 2nd ed., edited by Wolinsky, I. (Boca Raton, FL: CRC Press), pp. 257–275.

Stump, C.S., Houtkooper, L.B., Hewitt M.J., Going, S.B. and Lohman T.G., 1988, Bioelectric Impedance variability with hypohydration and exercise. *Medicine and Science in Sports and Exercise*, **20**, p. S82.

Takeuchi, K., Murata, K., Funaki, K., Fujita, I., Hayakawa, Y. and Morita, H., 2000, Bioelectrical impedance analysis in the clinical management of a pregnant woman undergoing dialysis. *Journal of Perinatal Medicine*, **28**, pp. 228–231.

Tatara, T. and Tsuzaki, K., 1998, Segmental bioelectrical impedance analysis improves the prediction for extracellular water volume changes during abdominal surgery. *Critical Care Medicine*, **26**, pp. 470–476.

Thomas, B.J., Cornish, B.H., Ward, L.C. and Patterson, M.A., 1998, A comparison of segmental and wrist-to-ankle methodologies of bioimpedance analysis. *Applied Radiation and Isotopes*, **49**, pp. 477–478.

Thomas, B.J., Cornish, B.H., Ward, L.C. and Jacobs, A., 1999, Bioimpedance: is it a predictor of true water volume? *Annals of the New York Academy of Sciences*, **873**, pp. 89–93.

Thompson, D.L., Thompson, W.R., Prestridge, T.J., Bailey, J.G., Bean, M.H., Brown, S.P. and McDaniel, J.B., 1991, Effects of hydration and dehydration on body composition analysis: a comparative study of bioelectric impedance analysis and hydrodensitometry. Journal of Sports Medicine and Physical Fitness, **31**, pp. 565–570.

van Marken Lichtenbelt, W.D., Westerterp, K.R., Wouters, L. and Luijendijk, S.C, 1994, Validation of bioelectrical-impedance measurements as a method to estimate body-water compartments. *American Journal of Clinical Nutrition*, **60**, pp. 159–166.

Vehrs, P., Morrow, J.R. Jr. and Butte, N., 1998, Reliability and concurrent validity of Futrex and bioelectrical impedance. *International Journal of Sports Medicine*, **19**, pp. 560–566.

Ward, L.C., Elia, M. and Cornish, B.H., 1998, Potential errors in the application of mixture theory to multifrequency bioelectrical impedance analysis. *Physiological Measurement*, **19**, pp. 53–60.

Zhu, F., Schneditz, D., Kaufman, A.M. and Levin, N.W., 2000, Estimation of body fluid changes during peritoneal dialysis by segmental bioimpedance analysis. *Kidney International*, **57**, pp. 299–306.

CHAPTER SEVEN

Anthropometric Measurements in Zambian Children

Zaida Cordero-MacIntyre[1], Rebecca Duran[1],
Shiva Metgalchi [1], Maribet Rivera[1], Gail Ormsby[2],
[1]Loma Linda University, Nutrition Department SPH, Loma Linda,
CA 92354, USA.
[2]Adventist Development and Relief Agency (ADRA) Australia.

1. INTRODUCTION

For field studies the use of anthropometry to assess body composition in adults and children is an ideal method. The basic equipment needed includes a hanging horizontal balance scale, a measuring tape and calipers and offer the advantage of portability and low cost. Specific equations to predict percentage body fat have been developed for groups of differing age, sex and ethnicity. The precision of this method is good (Lohman *et al.*, 1992). The standard errors of estimate are reported to be 3.6–3.9% fat for the triceps plus subscapular sites and 3.8% fat for the triceps plus calf sites (Slaughter *et al.*, 1988). Training of personnel is straightforward and the equipment is maintenance free (Ellis, 2001).

Bioelectrical impedance analysis (BIA) is another potentially convenient method for field use. The equipment is portable and relatively inexpensive. In the field, the equipment can work using a solar-charged or vehicle-charged deep-cell 12-volt battery or an electrical generator. This method is based on the principle that BIA estimates body composition based on the different conductive properties of different body tissues at various electrical frequencies. Tissues that are high in water and electrolyte content like muscle, blood and cerebrospinal fluid are highly conductive, while tissues like fat and bone are resistive to electrical conductivity. These properties of tissue conductivity, or more specifically impedance (total opposition to electrical current) are used to assess body composition. An applied current follows the path of least resistance, and charge may be temporarily stored in cell membranes as capacitance.

Impedance is a function of resistance and reactance according to the formula:

$$Z = (R^2 + Xc^2)^{0.5} \qquad\qquad (1)$$

Where Z = impedance in Ω
R = resistance in Ω
Xc = capacitance in Ω

Because resistance is much greater than capacitance at the standard 50kHz frequency, it largely determines the magnitude of the impedance, and Z and R are sometimes used interchangeably.   In a substance of uniform composition, resistance is proportional to the square of the path length and inversely proportional to the volume. While the living body is more complex, the volume of the different tissues can be deduced by measuring their combined resistances. (Ellis, 2001; Roche *et al.*, 1996).   In practice, once the resistance has been established, a BIA analyzer uses in-built prediction equations for estimating %fat which take account of stature, mass, age and sex, and activity level.

Jurimäe *et al.* (1998) studied the differences in assessing percent body fat (% BF) in pre-pubertal White European children using six different regression equations and BIA (Bodystat-500, Bodystat Ltd. Isle of Man. U.K.). Their results showed that for this group of children the Boileau and Houtkooper equations were the most appropriate.

Caballero *et al.* (1998) developed instruments to assess body fat in American-Indian children. The authors used isotopic dilution as a reference standard. A specific equation was developed on the basis of body weight, triceps and subscapular skinfold and BIA (Valhalla Scientific Model 1990B. San Diego, CA. U.S.A.).

Lohman *et al.* (1999) used this equation as well as BIA to assess body fatness in school-age children.   Their results showed that this group of children had excessive body fat.

Leman *et al.* (2003) validated the use of BIA (01Q; RJL Systems Inc., Clinton Township, MI) to assess body composition in Nigerian children and adults. The authors used total body water (TBW) determinations to develop prediction equations for this group of children and adults.

Vienna *et al.* (1998) used BIA (101/Quantum, AKERN s.r. Florence, Italy) and anthropometric measurements of height, weight and body mass index (BMI) in Ecuadorian children of African ancestry. Their findings showed BIA yielded much more information than that of the limited anthropometric measurements taken. BIA proved useful in determining body composition in school age children of different socioeconomic status and thus estimating the risk of malnutrition. Regardless of socioeconomic status, females had more fat and less total body water than males, and children of higher socioeconomic status had less fat than those of lower socioeconomic status.

VanderJagt *et al.* (2000) used BIA (Quantum, RJL, Inc. Clinton Township, MD) to assess body composition of children with sickle cell anemia in northern Nigeria. Their study showed that BIA was a useful method in assessing body composition in these children considering the dryness and high temperatures that prevail in this region of sub-Saharan Africa.

The intention of this study, therefore, was to compare the body composition results obtained by BIA-TANITA® TBF 300A to those of three prediction equations in African children in Zambia. We hypothesised that there would be no significant difference between the body composition results obtained using anthropometric measurements and BIA-TANITA® in this group of children. To our knowledge this was the first study to use BIA-TANITA® TBF 300A in Zambian children.

2. METHODS

2.1 Subjects

This investigation involved children from five rural villages (*Chaola, Chisunga, Fusi, Katambo,* and *Mohenga*), in the Mwami catchment area of the Eastern Province, Zambia, in conjunction with a Child Survival Project funded by the United States Agency for International Development (USAID).

The sample for this particular study was drawn from a follow-up cohort study of Mwami catchment children. During the follow-up process of conducting a dietary and anthropometric assessment, the children's mothers were interviewed along with their children. Fifty children were included in this study, 23 females and 26 males. The gender for one of the children was not recorded. However, since the results of this study are of pooled data of males and females the child for whom gender is missing was included in the data analysis.

This study was conducted in primitive conditions that are the environment in which the Child Survival Project functions. Measurements were conducted inside a rural, mud, straw-thatched village hut. The temperature was hot at an estimated 27-30°C dry heat during the months of August-September 1999. Under these circumstances we had no control over the environmental conditions of humidity or temperature.

Several authors have reported the effects of ambient temperature on BIA readings. Caton *et al.* (1988) showed that ambient temperature can influence BIA readings. Skin temperature and consequently blood flow change with temperature. It has been reported that temperatures of 35°C can produce lower readings for fat mass and higher fat free mass, as compared with lower ambient temperatures. Gudivaka *et al.* (1996) had similar findings when he subjected his subjects to a temperature of 35.8°C. Liang *et al.* (2000) also concluded that a decrease or an increase in ambient temperature affects both body temperature and skin blood flow. Consequently both affect BIA readings.

Buono *et al.* (2004) studied the effects of different temperatures (15, 20, 25, 30 and 35°C) on skin blood flow (SBF) and BIA results. These authors showed that BIA results are not significantly affected by temperatures of 20 and 25°C. However at the higher temperatures resistance was inversely related to skin blood flow, and this, rather than the electrode-skin interface, affected BIA output.

2.2 Measurements

To standardize measurement procedures, the ADRA personnel and Loma Linda University postgraduate students participating in this study were trained to perform anthropometric and BIA measurements by Loma Linda University faculty members.

2.3 Anthropometry

Measurements were made according to the protocols of Lohman *et al.* (1988), with the exception that measures were made on the left side of the body. Subjects wore light-weight clothing, and had bare feet. Standing height measurements were taken using metal tape fixed to a vertical board. Measurements were taken to the nearest 0.1 cm. A Broca plane was used to mark the child's height on the measuring tape. (Perspective Enterprises. Portage, Michigan. USA).

Skinfold measurements were taken to the nearest 1 mm using Lange skinfold calipers (Cambridge Scientific Industries, Inc. Cambridge, Maryland, USA). Measurements were taken while the subjects stood in a relaxed position wearing minimal clothing. Each measurement was repeated in a set of two trials and the mean of these measurements was used for calculations. The following skinfolds were measured: Triceps: Vertical fold raised midway between the olecranon and acromion process on the posterior of the brachium. Subscapular: Measured 1 cm below the inferior angle of the left scapular inclined downwards and laterally in the natural cleavage of the skin. Medial calf: Vertical skinfold raised on the medial side of the left calf just below the level of the maximal calf girth.

For all these measurements, the children's mothers assisted in removing excess clothing from the children so that the children were wearing minimal clothing when measured. The mothers also assisted in gaining the children's cooperation for the measurements while the trained students and ADRA personnel performed the procedures. Three different equations were used to predict children's body composition. Houtkooper's Equation (2) was used to calculate fat free mass (FFM), while Lohman's (3) and Slaughter's (4) Equations were used to calculate percent fat (%Fat).

Houtkooper's Equation
  For males and females $Kg = 0.61\ S^2/R + 0.25\ wt + 1.31$ (2)

  Where   S = Standing height (cm)
        R = Resistance (ohms)
        Wt = Kg.
        (The R value given by the BIA Tanita-BIA output).

Lohman's Equation
  Males: % Fat = 1.21 $(\sum SF) - 0.008 (\sum SF)^2 + I_M$
  Females: % Fat = 1.33 $(\sum SF) - 0.013 (\sum SF)^2 + 2.0$ (Females)   (3)

  Where $\sum SF$ = sum of Triceps and Subscapular skinfolds,
            I_M = is an intercept which varies with maturation level and racial
            group for males. We used (-3.5) for prepubescent black children
            (Lohman, 1992).

Slaughter's Equation
  Females % Fat= 0.610 (triceps and calf skinfold) + 5.0
  Males % Fat = 0.735 (triceps and calf skinfold) + 1.0   (4)

2.4 Bioelectrical Impedance

Leg to leg BIA-TANITA® TBF 300A (Tokyo, Japan) was used to assess the
children's percent fat (% Fat), Kilograms (kg) of fat free mass (FFM), and total
body mass (TBM). The BIA-TANITA® TBF 300A is a portable device that
resembles a bathroom scale; it requires a horizontal, smooth hard surface. In this
study, the device was set up inside a rural, mud, straw-thatched village hut. The
subjects (wearing minimal clothing) wiped their feet and then stood upon the
footplate electrodes that are located on top of the horizontal surface of the
apparatus. Subjects stood in a relaxed position while the BIA measurement was
taken. Two readings were taken for each subject. The mean of both measurements
was used in the data analysis.
The results of BIA-TANITA® TBF 300A were compared to those obtained from
Houtkooper (FFM), Lohman and Slaughter's (%Fat) equations.

3. DATA ANALYSIS

Descriptive statistical analysis was done using SPSS 10.0 version (SPSS Inc.
Chicago Ill). Assumptions were met according to Pearson's correlation (normal
distribution, equal variances, homogeneity, linearity between the two variables),
and the accuracy of between-person variation was evaluated.
    Pearson's correlation was used to test for a linear relationship between the
results for FFM obtained using Houtkooper's equation and BIA-TANITA® TBF
300A. The same was done with the results for %Fat using the Lohman, Slaughter
Equations and BIA results. Bland-Altman plots (Bland, 1986) were used as an
alternate method of analysis. The mean of two measurements was used in the
statistical analysis. Due to the small number of subjects, analysis was done using
pooled data of both males and females.

4. RESULTS

Table 1 shows the subjects characteristics. Results are presented as minimum, maximum, mean and standard deviation. Table 2 shows the Pearson correlation results

Table 1. Physical Characteristics of the Study Subjects

Variables	N	Min	Max	Mean	Std. Deviation
Height (cm)	50	91.40	140.10	107.99	7.99
Weight (kg)	50	12.60	33.50	17.79	3.33
Age (Month)	50	60	98	77.28	8.27
Gender					
Female	23	N/A	N/A	N/A	N/A
Male	26				
Missing	1				

N/A = Not Applicable

Table 2. Body Composition Correlations Between Body Composition Equations and BIA Tanita.

Equations	N	Mean ± SD	*r* value	*P* value
Houtkooper's FFM (kg) vs (TANITA)	50	16.16 ± 3.46 15.13 ± 2.90	0.79	<0.001
Lohman's % Fat vs (TANITA)	50	13.95 ± 5.03 15.00 ± 5.81	−0.36	0.011
Slaughter's % Fat vs (TANITA)	50	15.02 ± 3.17 15.00 ± 5.81	−0.10	0.47
Lohman's % Fat vs Slaughter's % Fat	5	13.95 ± 5.03 15.02 ± 3.17	0.76	0.0005
Houtkooper's FFM (%) vs (TANITA)	50 500	9.00 ± 0.14 15.00 ± 5.81	0.26	0.86

There was a significant positive correlation (r = 0.79, p < 0.001) between Houtkooper's Equation and BIA-TANITA® TBF 300A for FFM.

However, we are aware that the Houtkooper's Equation contains the R value (resistance) in the equation. This value is given by BIA-TANITA® TBF 300A.

There was a significant negative correlation (r = −0.36, p = 0.011) between Lohman's Equation for %Fat and BIA-TANITA® TBF 300A. A significant negative correlation was observed (r = −0.10, p = 0.47) between Slaughter's %Fat and BIA-TANITA® TBF 300A. However, a significant positive correlation was observed between Lohman's %Fat and Slaughter's %Fat (r = 0.76, p = < 0.0005). When we derived %Fat from Houtkooper's Equation, we found a non-significant correlation (r = 0.26, p = 0.86) between this equation and BIA-TANITA® TBF 300A.

Figure 1 shows the results of a Bland-Altman plot for FFM calculated using Houtkooper's Equation and FFM measured using BIA-TANITA® TBF 300A. Many children were found to have parasitic infections and the degree of infection may have influenced their total body mass. A couple of children were found to be heavier than the majority of the group as indicated by the expected body composition by BIA-TANITA® TBF 300A calculations for height and age, this may explain the outliers.

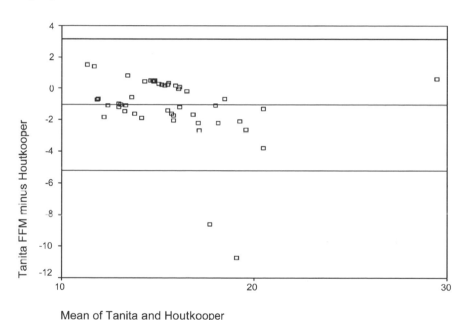

Mean of Tanita and Houtkooper

Figure 1. Bland-Altman plot showing the mean of the TANITA® and Houtkooper's equation (FFM) kg. and the difference between the two means

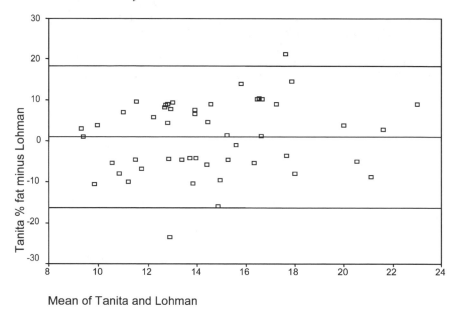

Mean of Tanita and Lohman

Figure 2. Bland-Altman plots showing the mean of the TANITA® and Lohman's (% Fat) equation and the difference between the two means

Figure 2 shows the results of Bland-Altman plot for % Fat using Lohman's equation and %Fat measured using BIA-TANITA® TBF 300A.

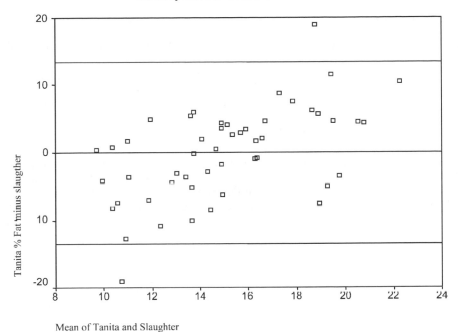

Mean of Tanita and Slaughter

Figure 3 Bland-Altman plot showing the mean of the TANITA® and Slaughter's (%Fat) equation and the difference between the two means

Figure 3 shows the results of Bland-Altman plot for %Fat using Slaughter's equation and %Fat measured using BIA-TANITA® TBF 300A.
These last two graphs show a large variability in %fat between BIA-TANITA® TBF 300A and Lohman's (±18%fat) and Slaughter's (±14%fat) equations.
The combination of the correlation results and the Bland-Altman plots show us that there is no agreement between methods.

5. DISCUSSION

The use of leg-to-leg BIA-TANITA® TBF 300A and different BIA and anthropometric equations to assess FFM and %Fat in children were used in this study in order to have a more efficient body composition tool for evaluating children in Zambia.

Although the results for FFM obtained using Houtkooper's Equation showed a significant correlation with those obtained with BIA-TANITA® TBF 300A, Bland-Altman plots show that there was large variability (±4 kg for children whose mean weight is 18 kg.

Our results were not in agreement with those obtained by other investigators. Jurimäe *et al.* (1998) reported results of their study conducted in 9-11 year old white children. In that study the authors compared BIA using Bodystat-500 to Houtkooper's Equation and concluded that this was an appropriate equation for use in this group of children.

The results for %fat that we obtained using Lohman's, Slaughter's and Houtkooper's equations were not comparable to those obtained with BIA-TANITA® TBF 300A as shown by the correlation results and Bland-Altman plots.

Besides temperature, other factors such as age, sex, growth, body type, and body weight may affect BIA results. Lohman *et al.* (1992) analyzed the effect of age and sex on hydration of Fat Free Body (FFB). His findings showed that with age, children and youth of both sexes experienced a decrease in hydration status. Males experienced a decrease of 3.6% and in females a decrease of 3-6%. In a prepubertal sample, these authors found the water content of the FFB to be 75% as compared to 72.5% in adults. These changes in hydration may influence BIA readings, giving lower readings for FM.

Bunc *et al.* (2000) showed that total body water (TBW) estimated by BIA (2000 M, Data Input, Germany) was higher in children than adults. The children's values decreased with age. Therefore, this will have an effect on FFM readings.

Phyllips *et al.* (2003) reported that BIA (101, RJL, Clinton Township, MI), provided accurate measurements for FFM and %FM in adolescent girls who were followed from September 1990 to June 1993

Jurimae *et al.* (2000) analyzed the effect of different somatotypes in 9-11 year old Caucasian children. Their findings showed that the mesomorphic type (robust and muscular) was the type that most influenced body resistance in both prepubescent boys and girls.

Ward *et al.* (2000) reported that BIA results may be affected by different population's body type or by ethnicity. It is not clear if the differences were due to body geometry or to electrical properties. What was clear was that body impedance exhibited population specific characteristics.

Schoeller *et al.* (2000), in their BIA study of African-Americans and Caucasians, found that excess body weight was the variable that influenced their results and not race.

All these studies point out the effect of other variables that may have influence on BIA results, which might have been the case in our study

At the time of this study (1999), there was a paucity of published work on the use of BIA measurement among general rural populations in Africa. This was one of the first studies to compare the use of BIA for body composition along with more traditional methods of anthropometric measurement in a rural setting. At the time of the study, no BIA reference data were available for children. The study was an early attempt to assess if BIA methods would work in these difficult situations. While we did expect to validate the skinfold prediction equations using BIA-TANITA® TBF 300A, due to the working conditions in the rural field, we did not reach the expected outcome. All of these factors resulted in our being unable to have confidence in the accuracy of our BIA measurements.

To our knowledge this is the first study that attempted to compare the results of the Houtkooper, Lohman and Slaughter's equations with those of leg-to-leg BIA-TANITA® TBF 300A, in Zambian children. Other investigators used other BIA devices to measure body composition in urban African children. These measured lower and upper body composition, while the device we used measured lower body composition. This fact may have also influenced the outcome of our results.

While the random error is high, the bias is small, so that using any of these methods to assess group means in a similar population may yield similar results. However, evaluation of individual levels of fatness can often be more important in making decisions for therapeutic interventions. In public health we are constantly faced with the choices for nutrition interventions based on the findings for each individual.

6. CONCLUSION

In summary, we suspect that due to the environmental conditions under which this study was conducted, the use of BIA-TANITA® TBF 300A for field use in this group of children was not validated. Therefore, we feel that the use of skinfolds to assess body composition in the rural setting of Zambia is still the preferred instrument for the field where it is impossible to control temperature, humidity and other factors.

In order to validate the use of BIA-TANITA® TBF 300A for field use in this group of Zambian children more studies are needed under controlled conditions of a laboratory setting to estimate the contribution of the separate errors which appear to have confounded the primary study aim.

Acknowledgements

The authors wish to express their gratitude to staff of ADRA Zambia and the Mwami Hospital who helped make this study possible. Further, appreciation is expressed to TANITA® who assisted with the provision of equipment for the study.

REFERENCES

Bland, J.M. and Altman, D.G., 1986, Statistical methods for assessing agreement between two methods of clinical measurement. *The Lancet*, 1, pp. 307-310.

Bunc, V., *et al.*, 2000, Estimation of body composition by multifrequency bioimpedance measurement in children. In *In vivo body composition studies*, edited by Yasumura, S., Wang, J., Pierson, R.N. (New York: The New York Academy of Sciences), pp. 203-204.

Buono, M.J., *et al.*, 2004, The effect of ambient air temperature on whole-body bioelectrical impedance. *Physiol Meas*, 25, pp. 119-123.

Caballero, B., *et al.*, 1998, Pathways: A school-based program for the primary prevention of obesity in American Indian children. *Journal of Nutrition and Biochemistry*, 9, pp. 535-543.

Caton, JR., *et al.*, 1988, Body composition analysis by bioelectrical impedance: effect of skin temperature. *Med Sci Sports Exerc*, 20, pp. 489-91.

Ellis, K.J., 2001, Selected body composition methods can be used in field studies. *Journal of Nutrition*, 131, pp. 1589S-1595S.

Gudivaka R., *et al.*, 1996, Effect of skin temperature on multifrequency bioelectrical impedance analysis. *J Appl Physiol*, 81, pp. 838-45.

Houtkooper, L.B., *et al.*, 1996, Why bioelectric impedance analysis should be used for estimating adiposity. Bioelectric Impedance Technology Assessment Conference. *Am J Clin Nutr*, 64, pp. 4365-4485.

Jurimäe, T., *et al.*, 2000, Relationships between bioelectric resistance and somatotype in 9- to 11-year-old children. In *In vivo body composition studies*, edited by Yasumura, S., Wang, J., Pierson, R.N. (New York: The New York Academy of Sciences), pp. 187-192.

Jurimäe, T., *et al.*, 1998, Assessment of body composition in 9 to 11 year old children by skinfold thickness measurements and bioelectrical impedance analysis: comparison of different regression equations. *Medicina Dello* Sport, 51, pp. 341-347.

Leman, C.R., *et al.*, 2003, Body composition of children in south-western Nigeria: validation of bio-electrical impedance analysis. *Annals of Tropical Pediatrics*, 23, pp. 61-67.

Liang, M.T., *et al.*, 2000, Skin temperature and skin blood flow affect bioelectric impedance study of female fat-free mass. *Med Sci Sports Exerc*, 32, pp. 221-227.

Lohman, T.G., *et al.*, 1984, Bone mineral content measurements and their relation to body density in children, youth, and adults. *Human Biology*, 56, pp. 667-679.

Lohman, T.G., *et al.*, 1988, *Anthropometric standardization reference manual.* (Human Kinetics: Champaign, IL).

Lohman, T.G., 1992, Advances in body composition assessment. In Current Issues in Exercise Science Monograph Number 3. (Champaign: Human Kinetics Publishers), p. 65-74.

Lohman, T.G., *et al.*, 1999, Body composition assessment in American Indian children. *American Journal of Clinical Nutrition*, 69, pp. 764S-766S.

Lohman, T.G., *et al.*, 2000, Estimation of body fat from anthropometry and bioelectrical impedance in Native American children. *International Journal of Obesity*, 24, pp. 1-7.

Malina, R.M., 1973, Biological substrata. In *Comparative studies of blacks and whites in the United States*, edited by Miller K.S. and Dreger R.M. (New York: Seminar Press), pp. 53-123.

Malina, R.M., 1996, Regional body composition: age, sex, and ethnic variation. In *Human Body Composition*, edited by Roche, A.F., *et al.* (Champaign: Human Kinetics), pp. 217-255.

Phillips, S.M., *et al.*, 2003, A Longitudinal Comparison of Body Composition by Total Body Water and Bioelectrical Impedance in Adolescent Girls. *J Nutr*, 133, pp. 1419-1425.

Roche, A.F., *et al.*, 1996, *Human Body Composition*, edited by Roche, A.F., *et al.* (Champaign: Human Kinetics Publishers), p. 80.

Schoeller, D.A. and Luke A., 2000, Bioelectrical impedance analysis prediction equations differ between African Americans and Caucasians, but it is not clear why. In *In vivo body composition studies*, edited by Yasumura, S., Wang, J., Pierson, R.N. (New York: The New York Academy of Sciences), pp. 225-226.

Slaughter M.H., *et al.*, 1988, Skinfold equations for estimation of body fatness in children and youth. Hum Biol, 60, pp. 709-723.

VanderJagt, D.J., *et al.*, 2000, Bioelectrical impedance analysis of the body composition of Nigerian children with sickle cell disease. *Journal of Tropical Pediatrics*, 46, pp. 67-72.

Vienna, A., *et al.*, 1998, Bioelectrical impedance analysis and anthropometry in Ecuadorian children of African ancestry. *Collegium Antropologicum*, 22, pp. 433-446.

Wagner, D.R. and Heyward, V.H., 2000, Measures of body composition in blacks and whites: a comparative review. *The American Journal of Clinical Nutrition*, 71, pp. 1392-1402.

Ward, L.C., *et al.*, 2000, Association between ethnicity, body mass index, and bioelectrical impedance. In *In vivo body composition studies*, edited by Yasumura, S., Wang, J., Pierson, R.N. (New York: The New York Academy of Sciences), pp. 199-202.

CHAPTER EIGHT

Pubertal Maturation, Hormonal Levels and Body Composition in Elite Gymnasts

Panagiota Klentrou[1], Andreas D. Flouris[2] and Michael Plyley[1]
[1] Department of Physical Education and Kinesiology, Brock University, Ontario, Canada
[2] Environmental Ergonomics Laboratory, School of Health and Human Performance, Dalhousie University, Nova Scotia, Canada

1. INTRODUCTION

The increasing involvement of young children in intense physical training over the past decades has generated concerns as to its potential effects on children's growth, maturation, and reproductive function (Mansfield and Emans, 1993). One of the sports liable for such concerns is gymnastics which, compared to other sports, entails initiation and intense physical training at a very young age. Reports in the literature describe weekly training volumes of young prepubertal and/or early pubertal gymnasts as long as 36 hours (Bass *et al.*, 1998). It is not surprising, therefore, that some authors have described elite female gymnasts as at risk for short stature and delayed maturation (Frisch *et al.*, 1981; Theintz *et al.*, 1993; Bass *et al.*, 2000). Elite male gymnasts have been found generally advanced in pubertal development for their age, but are generally short in stature (Buckler and Brodie, 1977). However, it seems reasonable that these results may be attributed to selection bias rather than gymnastics training *per se*.

The present investigation compared the sexual maturation, hormonal levels, and body fat of elite Canadian gymnasts to age-matched controls. It was hypothesized that intense training would have resulted to a delay in physical growth and sexual maturation in young male and female gymnasts. Sexual maturation was self-assessed (Tanner scale) and relative body fat was measured by BIA. Salivary testosterone was measured in males, and plasma estradiol-17-β and progesterone were measured in females. For menstruating girls, age at menarche was reported and blood draws were performed twice (in follicular and luteal phase). Results showed that although the intense training by the young male gymnasts resulted in decreased body fat, there was no apparent alteration in the timing and extent of their physical and pubertal development. Delayed menarche, reduced physical and sexual maturation, and low body fat were found in female gymnasts.

2. EFFECTS OF PHYSICAL TRAINING ON HUMAN DEVELOPMENT

Puberty in humans is characterized by large hormonal changes resulting in both physical maturation (i.e., skeletal development and growth) and sexual maturation (growth of pubic hair, and development in the genitalia). Intense training has been found to delay the onset of puberty in females by altering normal hormonal development (Theintz *et al.*, 1994; Theintz, 1994). This results in delayed pubertal onset, delayed age at first menarche, and failure to develop mature skeletal structure (Malina, 1983). In males, despite evidence that physical activity can also result in hormonal changes, there have been few studies that actually examined the relationship between training and the onset of puberty.

Puberty is triggered in males by an increase in secretion of gonadotropin releasing hormone (GnRH) from the hypothalamus. GnRH acts on the pituitary stimulating the release of luteinizing hormone (LH) and follicle-stimulating hormone (FSH) that cause increased release of testosterone from the gonads. Testosterone continues to increase throughout puberty until it reaches adult levels and plays a key role in both physical and sexual maturation (Rilling *et al.*, 1996). Studies on elite teams in popular sports (i.e., football and baseball) have provided evidence that physical activity causes early onset of puberty in males (Hale, 1956; Cacciari *et al.*, 1990). However, it seems reasonable that these results are subject to sample bias as the sports that were studied favour athletes who are more physically mature than other athletes of their age. Young males who participate in individual sports often train from an early age providing a better model to examine the effects of physical activity on puberty. Studies on adults report that endurance training results in decreased resting levels of circulating serum testosterone (Hackney, 1989; Elias and Wilson, 1993). In addition, recent data from 20-year old elite rowers suggested that, although a 22% increase in training intensity results in a 9% increase in testosterone, the later is significantly reduced to pre-training levels following two weeks of tapering (Maetsu *et al.*, 2003). These results are in agreement with a previous study (Carli *et al.*, 1983) demonstrating that 43 weeks of swim training significantly decreases testosterone to below pre-training levels in pubertal athletes. In line with these results, studies using gymnasts have found evidence of lowered testosterone following only three days of training (Rich *et al.*, 1992).

Growth during the prepubertal period is predominantly regulated by growth hormone, insulin-like growth factor-I (IGF-1), and thyroxine (Clark and Rogol, 1996). Although exercise has been shown to stimulate growth hormone release in adolescent boys (Zakas *et al.*, 1994), reports in the literature suggest that male gymnasts undertaking intensive training demonstrate reduced height and skeletal maturity relative to chronological age (Keller and Frohner, 1989). However, the lack of a difference in growth rates, IGF-1, and diet between young male gymnasts and controls over a 10 to 18-month period indicates that the short stature found in gymnasts may be due to selection bias rather than gymnastics training *per se* (Daly *et al.*, 1998; Daly *et al.*, 2000).

Female athletes who begin training at a young age often have a delay in menarche (Frisch *et al.*, 1981). The time at which athletic training was initiated has been implicated as a factor in delayed menarche since intense training prior to puberty may alter hypothalamic-pituitary function. Body composition has also been used to explain the delayed menstruation and menstrual cycle irregularities among elite athletes. Although not supported by current opinion, an association between menstrual regularity and the level of body fat has been theorized, suggesting that a level of at least 17% body fat is critical for the onset of menstruation and the maintenance of a normal cycle (Frisch and McArthur, 1974). On the other hand, more recent studies have suggested that delayed menarche may be due to genetic factors, and that girls who mature later often self-select sports which require high relative strength and small bodies, such as gymnastics (Loucks, 1995; Baxter-Jones and Helms, 1996). Indeed, 37% of the variance in the age at menarche has been found to be attributed to additive genetic factors (Kaprio *et al.*, 1995). In addition, the correlation between the additive genetic effects on age at menarche and body mass index was 0.57, indicating that a substantial proportion of genetic effects are common (Kaprio *et al.*, 1995). Further, it has been suggested that heavy training in gymnastics (>18 hr/wk), starting before puberty and maintained throughout puberty, can alter growth rate of females to such an extent that full adult height will not be reached (Theintz *et al.*, 1993). These authors suggested that the prolonged inhibition of the hypothalamic-pituitary-gonadal axis by exercise, together with or because of the metabolic effects of dieting, is responsible for these results (Theintz *et al.*, 1993). However, the exact underlying mechanisms generating the adverse effects of heavy gymnastics training on females' growth have not been established hitherto.

The purpose of this research is to evaluate the effect of intense training during somatic growth and skeletal development on sexual maturation, hormonal levels, and body fat in elite male and female gymnasts as compared to an age-matched non-training control group. It was hypothesized that following several years of training the increased energy expenditure by elite gymnasts would have resulted to lower levels of sex hormones in young gymnasts than other children of their age leading to a delay in physical growth characteristics and markers of sexual maturation.

3. METHODOLOGY

3.1 Participants

Twenty-one male elite gymnasts (M-GYM; 13.3 ± 0.3y) and 18 female gymnasts (F-GYM; 13.4 ± 0.2y), were recruited from competitive gymnastics clubs in Ontario, Canada. In order to qualify for the experimental group, gymnasts had to be competing at the provincial level or above and training at least 15 hours per week. Twenty-two age-matched boys (M-CON; 13.5 ± 0.3 years) and 24 age-matched girls (F-CON; 13.9 ± 0.1y) were recruited from recreational sport classes to participate as non-training controls. These children played sports not more than 2 hours per week. All participants and their parents were informed of any risks that might result from participation and informed consent was obtained before the initiation of the study. The experimental protocol was approved by the Brock University Research Ethics Board.

3.2 Protocol

Each participant was tested on one occasion at his/her gymnastic or recreational club. The testing was completed by the same investigators in the late afternoon (around 1800h) before the participants' regular activity session. During testing individuals had their physical characteristics measured, completed a physical activity questionnaire, and completed a self-assessment of pubertal stages using the Tanner scale (Tanner, 1962). All male subjects provided saliva samples to measure testosterone and all female subjects provided blood samples to measure estrogen and progesterone. Prior to all tests participants were reminded that all results were to be recorded anonymously.

3.3 Assessments

Age (accurate to 1 month) was recorded. Standing height was measured to the nearest 0.5 cm (Seca Stadiometer 208) with the participant's shoes off, feet together and head in the Frankfort horizontal plane. Body mass was precisely measured using an electronic scale (Tanita, TBF-521, Body Fat Monitor/Scale, Japan) to the nearest 0.1 kg with participant's shoes, sweaters, coats and jackets removed. Relative body fat (%BF) was evaluated using Bioelectrical Impedance Analysis (BIA) with a portable hand-held body composition monitoring unit (RJL Systems, MI, USA). During the measurement, the participants were asked to lie supine while two electrodes were placed at the top part of the right hand and foot. Since dehydration is considered a limitation of this method, participants consumed water *ad libitum* approximately 20 minutes prior to data collection. Further, participants were instructed to refrain from exercise as well as caffeine and other diuretic drinks on the day of the test. Fat free mass (FFM) was calculated using the input variables of body frame size, height, mass, and sex (Kyle *et al.*, 2001): FFM (kg) = $-4.104 + (0.518$ x height $(cm)^2$/resistance) $+ (0.231$ x mass (kg)) $+ (0.130$ x reactance) $+ (4.229$ x sex: men $= 1$, women $= 0$). Short- and long-term reproducibility of this method was reported as r = 0.999 for measurements taken in the same participant within one week, and 0.977 for repeat measurements up to one month (Kyle *et al.*, 2001). Recently, the validity and reliability of this method has been demonstrated successfully in children and adolescents (Okasora *et al.*, 1999; Sung *et al.*, 2001).

Training sessions and training hours per week were recorded. Pubertal maturation was self-reported using the pictures of Sexual Maturation Scale by Tanner (Tanner, 1962). Each participant went into a room by him/herself where they completed the self-assessment to reduce embarrassment. Once completed the self-assessment was put into a plain folder by the participant and handed directly to the researcher present to maintain anonymity. Despite the existence of limitations when assessing maturation in youth using self-report methods, self-assessment of sexual maturation has showed excellent agreement (kappa = 0.81 to 0.91) with paediatrician assessment (Duke *et al.*, 1980). Past and recent studies agree that the use of adolescent self-staging appears of value in studying puberty in adolescents (Duke *et al.*, 1980; Varona-Lopez *et al.*, 1988; Schall *et al.*, 2002).

3.4 Salivary Testosterone

One milliliter of unstimulated whole mixed saliva was collected from each male individual using cylinder-shape swabs placed in the mouth for 1 minute. After sampling, the swabs were placed directly into plastic tubes. The samples were then centrifuged before freezing. The centrifuged saliva was maintained at -20°C until being assayed. No preservative was used in the collection tubes, and all saliva samples were collected in the early evening (around 1800h). The participants were asked not to consume any food or drink, excluding water, for at least one hour prior to saliva collection. The saliva samples were analyzed in duplicate by a trained radioimmunoassay (RIA) technician experienced in saliva determinations. Testosterone (T) was quantified using a Coat-A-Count Testosterone kit (Diagnostic Products, Los Angeles CA) modified for saliva. In brief, the saliva was submitted to a double ether extraction then preceded to RIA. To accommodate saliva, the calibrators were diluted 1:20 and to further increase detectability at the low end of the curve, a 5 pg standard was added giving a range of 5 - 800 pg. For each assay tube, 200 uL of sample was pipetted into a polypropylene tube coated with antibody. One mL of 125I-labelled T was added and an extended incubation time of 22 hours was used, at room temperature. Following incubation, the tubes were decanted then counted for 60 sec in an LKB 1272 gamma counter (Wallac Oy, Turku Finland). The antibody used in the assay is highly specific for testosterone with <5% cross-reactivity with DHT. The samples were analyzed in two separate runs. The sensitivities of the two assays were calculated to be 5 pg and the intra-assay coefficients of variation were 14% and 6%, respectively, averaged across low, medium, and high pools. Following standard procedures (Gurd and Klentrou, 2003), the mean T concentration from the two duplicates of each saliva specimen was used for our statistical analysis throughout.

3.5 Determination of Menstrual Status, Estrogen and Progesterone

Menarcheal female participants reported their age at menarche onset and day 1 (i.e., the first day of flow) for their last three menstrual cycles. They were then instructed to report day 1 of their next cycle, after the initial study visit. Blood testing sessions were scheduled to correspond to each of the desired phases based on this information. For menstruating participants (n = 22), blood draws were performed twice, during the follicular phase (days 1-5), and during the luteal phase (days 19-22) of their next menstrual cycle. For the pre-menarcheal individuals (n = 20), blood was drawn prior to their initial testing session.

Intravenous blood samples were taken from the antecubital vein into a vacutainer. All blood samples were immediately centrifuged for 15 minutes at 1500g. The separated plasma was removed from the sample tube, placed in 2 ml microtubes, and stored in a -20°C freezer until analysis. Estradiol (17-β) and progesterone levels were analyzed using Immulite ® (Diagnostic Products Corporation, USA). The intra-assay and inter-assay coefficients of variation for the estradiol assay were 7.3% and 6.0%, respectively. The intra-assay and inter-assay coefficients of variation for the progesterone assay were 7.1% and 7.6%, respectively.

3.6 Data Analysis

One way analysis of variance (ANOVA) incorporating Bonferroni adjustment was used to determine differences between the GYM and CON groups for all of the variables tested. All statistical analyses were carried out using SPSS (version 10, SPSS Inc., Chicago, Illinois) statistical software package. The level of significance was set at $p < 0.05$.

4. RESULTS

Male and female gymnasts trained an average of 4.7 ± 0.4 sessions per week for a duration of 18.7 ± 1.4 hours per week, which was significantly ($p < 0.05$) greater than what was reported by the controls. Male gymnasts were shorter and lighter than the controls and had significantly ($p < 0.05$) lower relative body fat (Table 1). In terms of sexual maturation, there were no significant differences detected in testosterone levels between M-GYM and M-CON (Table 2). There were also no significant differences between male groups for either the genital development or pubic hair development (Table 2).

Table 1 Physical characteristics (mean±SE) in male (M-GYM) and female gymnasts (F-GYM), and male (M-COM) and female controls (CON).

	Males		Females	
	M-GYM (n=21)	**M-CON (n=22)**	**F-GYM (n=18)**	**F-CON (n=24)**
Height (cm)	155.2 ± 2.4	$163.7 \pm 2.1*$	155.3 ± 1.6	$162.4 \pm 1.4*$
Weight (kg)	48.3 ± 2.7	$58.1 \pm 1.5*$	41.2 ± 2.7	$58.5 \pm 2.4*$
Relative Body Fat (%)	$8.6 \pm 0.1*$	$13.9 \pm 0.7*$	$15.1 \pm 0.5*$	$23.8 \pm 1.6*$
Lean Body Mass (kg)	44.0 ± 1.5	$46.8 \pm 0.4*$	37.9 ± 0.9	$42.9 \pm 1.03*$

Note: * = $p < 0.05$ between GYM and CON within the same gender

Table 2 Physical characteristics (mean±SE) and sexual maturation for male gymnasts (M-GYM) and controls (M-CON).

	M-GYM (n=21)	**M-CON (n=24)**	**Δ Between Groups**
Testosterone (pg·ml^{-1})	43.3 ± 5.8	38.2 ± 4.1	$p=0.25$
Genital Development (Tanner stage)	3.2 ± 0.2	3.5 ± 0.1	$p=0.15$
Pubic Hair Development (Tanner stage)	3.5 ± 0.1	3.6 ± 0.1	$p=0.11$

Note: No significant differences between M-GYM and M-CON for any variable

Female gymnasts were significantly (p<0.05) smaller, and had significantly (p<0.05) lower body fat than female controls (Table 1). Twenty-one female controls and one gymnast were menarcheal, with the average age at menarche being 12 ± 0.1y for F-CON, and age 13 for the menarcheal gymnast. The mean plasma estradiol-17-β and progesterone of female participants are presented in Table 3. Estradiol-17-β and progesterone levels for the pre-menarcheal gymnasts were similar to those measured of the controls during the follicular phase, which were significantly lower (p<0.05) than those measured during the luteal phase (Table 3). Breast development was delayed in female gymnasts compared to controls (stages: 2.7 ± 0.2 vs 3.8 ± 0.1; p=0.04), while the difference between the groups in pubic hair development did not reach significance (stages: 3.1 ± 0.2 vs 4.0 ± 0.2; p=0.075).

Table 3 Resting Estradiol-17-β and Progesterone levels (mean±SD) in female gymnasts (F-GYM) and controls (F-CON) who had reached menarche.

	Follicular Phase		Luteal Phase		Δ
	F-GYM (n=1)	F-CON (n=21)	F-GYM (n=1)	F-CON (n=21)	Between Phases
Estradiol-17-β (pmol·L⁻¹)	74.7	75.3 ± 1.1	119.7	212.5 ± 51.3	p=0.005
Progesterone (nmol·L⁻¹)	0.8	1.1 ± 0.03	7.5	16.3 ± 0.7	p=0.016

5. DISCUSSION

The present investigation compared the sexual maturation, hormonal levels, and body fat levels of elite gymnasts to a non-training control group. The major finding was that training in young male gymnasts did not significantly change resting salivary testosterone or alter the onset of puberty as determined by self-assessment of pubertal stages. Additionally, while the male gymnasts had lower %BF than their control peers, no significant differences were detected in height, weight or lean body mass between the two groups. In contrast, a delayed menarche and breast development, a low %BF and lower estrogen and progesterone levels were found in the females engaged in rigorous gymnastics training.

5.1 Effects of training on physical maturation of male gymnasts

In the present study the male gymnasts were shorter and lighter than the age-matched controls. Previous studies also found male gymnasts to be shorter than controls but concluded that gymnasts were shorter as a result of selection, because of an advantage of shorter athletes in the sport, and not developmental delays (Keller and Frohner, 1989; Daly *et al.*, 2000). Others have found that, although shorter than average, gymnasts are advanced in pubertal development for their age (Buckler and Brodie, 1977).

While the present results differed from those of Buckler and Brodie (1977) in terms of height and weight, both studies show significantly lower relative body fat in trained gymnasts. Furthermore, the lower body fat percentage without a difference in weight suggests an increase in lean body mass in response to training in our participants. This supports the results of Elias and Wilson (1993) who found that exercise during puberty results in increases in lean body mass and decreases in body fat.

Previous studies have demonstrated both an increase in testosterone levels as a result of both short-term (Wilson *et al.*, 1981; Hackney, 1989; Cacciari *et al.*, 1990; Mero *et al.*, 1990) and long-term training (Fahey *et al.*, 1979; Malina, 1983). The cases where testosterone levels decreased following training involved prolonged endurance type exercise (i.e., swimming), except Rich *et al.* (1992), who found a decrease in testosterone following three days of gymnastics training. In this study there was, however, an observed return towards pre-exercise levels following only one day of rest suggesting that the observed changes may have only been transient. Recently, it has been reported that a 22% increase in training intensity caused a 9% increase in testosterone in 20 year old male rowers but further increases of training volume by 25% resulted in no further changes in testosterone (Maetsu *et al.*, 2003). The present study found no significant difference in resting testosterone levels between the gymnasts and the controls. These results are supported by Fahey *et al.* (1979) who found that, while resting testosterone levels increased with pubertal stage, there were no observed differences following maximal exercise. Moreover, Daly *et al.* (1998) found no difference in resting serum testosterone, IGF-1 and cortisol between peripubertal gymnasts and controls at any time during a 10-month period. The data thus far, in addition to the absence of significant correlation between resting salivary testosterone levels and training volume, suggest that gymnastics may cause transient or acute alterations in testosterone levels in males but these changes may not persist chronically.

There is relatively little literature on the effect of training on pubertal development as measured by pubertal stages. Intense training has been shown to delay pubertal onset (Larzon and Klinger, 1989). However, this research involved individual case studies and may not be relevant to the population as a whole. A different study suggested that gymnasts were two years delayed in specific pubertal markers and concluded that this was a result of gymnastics selecting boys who were smaller and more likely to mature later, regardless of activity levels (Hackney, 1989). This supports the present results suggesting that gymnastics training had no effect on the pubertal development of young male gymnasts.

5.2 Effects of training on physical maturation of female gymnasts

As reported elsewhere, girls who mature later often self-select sports which require a high relative strength and small bodies, such as gymnastics (Loucks, 1995; Baxter-Jones and Helms, 1996). The late maturing, pre-menarcheal body type favours athletic success in this type of activity. This observation is in accordance with the present data in that the female gymnasts were not only significantly leaner, but were also significantly shorter than controls. Additional factors that may influence growth potential of female gymnasts include psychological and/or emotional stress resulting from frequent competitions, altered social relationships with peers, as well as demanding parents and/or coaches (Tofler *et al.*, 1996).

Twenty-one female controls and only one gymnast were menarcheal. The age of menarche for the female gymnast who had achieved menarche was delayed compared to both the controls in this study and to the controls in other studies (Malina, 1973; Theintz *et al.*, 1993; Constantini and Warren, 1995). Previous studies have also suggested that in females the age of the onset of pubertal development varies greatly, and is influenced by nutrition, heredity, state of health, percentage of body fat, and other factors (Hamm, 1991; Lindholm *et al.*, 1994). As reported, while engaging in intensive training, many young elite female athletes, including gymnasts, have an unsatisfactory nutritional status and very low body fat (Frisch *et al.*, 1981; Calabrese *et al.*, 1983; Calabrese, 1985; Theintz *et al.*, 1994; Theintz, 1994). The female gymnasts in the present study had significantly lower body weight, and %BF values than the normal control participants. Accordingly, breast development was delayed in gymnasts compared to controls (stages: 2.7 ± 0.2 vs 3.8 ± 0.1), while the difference between the groups in pubic hair development did not reach significance (stages: 3.1 ± 0.2 vs 4.0 ± 0.2). Breast development was also delayed in ballet dancers, who showed little or no development (stage 1 or 2) at 13 years of age, as compared to normal children who usually have reached stage 4 at menarche (Warren, 1980).

Moreover, it has been hypothesized that when puberty is temporarily interrupted by exercise, skeletal age maturation often stalls as growth slows and sex steroids fall to low levels (Theintz *et al.*, 1994; Theintz, 1994). As expected, estradiol and progesterone levels for the pre-menarcheal gymnasts were similar to those measured of the controls during the follicular phase. The estradiol levels reported in our study for the controls did not reach adult levels for both follicular and luteal phase, but they are similar to those previously reported for adolescents (Apter *et al.*, 1987). The plasma progesterone levels reported in the present study for the female controls are in the same range as that reported for adults (Charkoudian and Johnson, 1999). Luteal phase estradiol and progesterone concentrations for the one menarcheal gymnast were lower than the concentrations measured in the controls. Previous studies have also reported lower levels of luteal phase estradiol and progesterone in teenage swimmers as compared to age-matched controls and adults leading to the suggestion that corpora lutea in the swimmers were not functioning properly (Bonen *et al.*, 1981).

6. CONCLUSIONS

Results from the present study demonstrate that a high volume of gymnastics training did not influence the onset of puberty in young male elite Canadian gymnasts. While training did result in lower relative body fat there were no differences in height, weight, lean body mass, resting salivary testosterone or pubertal development between the male gymnasts and age-matched controls. These findings suggest that gymnastics training in young males does not appear to have significant effects on their resting testosterone and sexual maturation. On the other hand, a delayed menarche, reduced physical and sexual maturation, and a low relative body fat were found in young females engaged in high volume gymnastics training. More research is required to investigate the long-term effects of gymnastics training on growth, maturation, and growth factors in elite adolescent athletes.

REFERENCES

Apter, D., Räisänen, I., Ylöstalo, P. R. V., 1987, Follicular growth in relation to serum hormonal patterns in adolescent compared with adult menstrual cycle. *Fertility and Sterility,* **47**, pp. 82-88.

Bass, S., Pearce, G., Bradney, M., Hendrich, E., Delmas, P.D., Harding, A., *et al.*, 1998, Exercise before puberty may confer residual benefits in bone density in adulthood: studies in active prepubertal and retired female gymnasts. *Journal of Bone and Mineral Research,* **13**, pp. 500-507.

Bass, S., Bradney, M., Pearce, G., Hendrich, E., Inge, K., Stuckey, S. *et al.*, 2000, Short stature and delayed puberty in gymnasts: influence of selection bias on leg length and the duration of training on trunk length. *Journal of Pediatrics,* **136**, pp. 149-55.

Baxter-Jones, A.D.G. and Helms, P.J., 1996, Effects of training at a young age: a review of the training of young athletes (TOYA) study. *Pedeatric Exercise Science,* **8**, pp. 310-327.

Bonen, A., Belcastro, A.N., Ling, W.Y., Simpson, A.A., 1981, Profiles of selected hormones during menstrual cycles of teenage athletes. *Journal of Applied Physiology,* **50**, pp. 545-551.

Buckler, J.M.H. and Brodie, D.A., 1977, Growth and maturity characteristics of schoolboy gymnasts. *Annals of Human Biology,* **4**, pp. 455-463.

Cacciari, E., Mazzanti, L., Tassinari, D., Bergamaschi, R., Magnani, C., Zappulla, F., *et al.*, 1990, Effects of sport (football) on growth: auxological, , anthropometric and hormonal aspects. *European Journal of Applied Physiology and Occupational Physiology,* **61**, pp. 149-58.

Calabrese, L.H., 1985, Nutritional and medical aspects of gymnastics. *Clinics in Sports Medicine,* **4**, pp. 23-30.

Calabrese, L.H., Kirkendall, D.T. and Floyd, M., 1983, Menstrual abnormalities, nutritional patterns, and body composition in female classical ballet dancers. *The Physician and Sportsmedicine,* **11**, pp. 86-98.

Carli, G., Martelli, G., Viti, A., Baldi, L., Bonifazi, M. and Lupo di Prisco, C., 1983, Modulation of hormone levels in male swimmers during training. In: Hollander A, Huijing P, de Groot D, editors. *Biomechanics in Medicine and Swimming.* (Champaign IL: Human Kinetics), pp. 33-40.

Charkoudian, N.C. and Johnson, J.M., 1999, Altered reflex control of cutaneous circulation by female sex steroids is independent of prostaglandins. *American Journal of Physiology,* **276**, pp. H1634- H1640.

Clark, P.A. and Rogol, A.D., 1996, Growth hormones and sex steroid interactions at puberty. *Endocrinology and Metabolism Clinics of North America,* **25,** pp. 665-681.

Constantini, N.W. and Warren, M.P., 1995, Menstrual dysfunction in swimmers: a distinct entity. *The Journal of Clinical Endocrinology and Metabolism,* **80,** pp. 2740-2744.

Daly, R.M., Rich, P.A. and Klein, R., 1998, Hormonal responses to physical training in high-level peripubertal male gymnasts. *European Journal of Applied Physiology,* **79,** pp. 74-81.

Daly, R.M., Rich, P.A., Klein, R. and Bass, S.L., 2000, Short stature in competitive prepubertal and early pubertal male gymnasts: the result of selection bias or intense training? *Journal of Pediatrics,* **137,** pp. 510-516.

Duke, P.M., Litt, I.F. and Gross, R.T., 1980, Adolescents' self-assessment of sexual maturation. *Pediatrics,* **66,** pp. 918-920.

Elias, A.N. and Wilson, A.F., 1993, Exercise and gonadal function. *Human Reproduction,* **8,** pp. 1747-1761.

Fahey, T.D., Valle-Zuris, A.D., Oehlsen, G., Trieb, M. and Seymour, J., 1979, Pubertal stage differences in hormonal and hematological responses to maximal exercise in males. *Journal of Applied Physiology,* **46,** pp. 823-827.

Frisch, R., Gotz-Welbergen, A.V. and McArthur, J.W., 1981, Delayed menarche and amenorrhea of college athletes in relation to age of onset of training. *Journal of the American Medical Association,* **246,** pp. 1559-1563.

Frisch, R.E. and McArthur, J.W., 1974, Menstrual cycles: Fatness as a determinant of minimum weight for height necessary for their maintenance or onset. *Science,* **185,** pp. 949-951.

Gurd, B. and Klentrou, P., 2003, Physical and pubertal development in young male gymnasts. *Journal of Applied Physiology,* **95,** pp. 1011-1015.

Hackney, A.C., 1989, Endurance training and testosterone levels. *Sports Medicine,* **8,** pp. 117-127.

Hale, C.J., 1956, Physiologic maturity of little league baseball players. *Research Quarterly,* **27,** pp. 276-284.

Hamm, T., 1991, Physiology of normal female bleeding. *NAACOG's Clinical Issues in Perinatal and Women's Health Nursing,* **2,** pp. 289-294.

Kaprio, J., Rimpela, A., Winter, T., Viken, R.J., Rimpela, M. and Rose, R.J., 1995, Common genetic influences on BMI and age at menarche. *Humman Biology,* **67,** pp. 739-753.

Keller, E. and Frohner, G., 1989, *Growth and development of boys with intensive training in gymnastics during puberty.* (Florence, Italy: Sereno Symposium Publishers).

Kyle, U.G., Gremion, G., Genton, L., Slosman, D.O., Golay, A. and Pichard, C., 2001, Physical activity and fat-free and fat mass by bioelectrical impedance in 3853 adults. *Medicine and Science in Sports and Exercise,* **33,** pp. 576-584.

Larzon, Z. and Klinger, B., 1989, *Does intensive sport indanger normal growth and development.* (Florence, Italy: Sereno Symposium Publishers).

Lindholm, C., Hagenfeldt, K. and Rangertz, B., 1994, Pubertal development in elite juvenile gymnasts: effects of physical training. *Acta Obstetricia et Gynecologica Scandinavica,* **73**, pp. 269-273.

Loucks, A.B., 1995, The reproductive system and physical activity in adolescents. In: Blimkie CJR, Bar-Or O, editors. *New Horizons in Pediatric Exercise Science.* (Champaign, IL.: Human Kinetics), pp. 27-37.

Maetsu, J., Jurimae, J. and Jurimae, T., 2003, Hormonal reactions during heavy training stress and following tapering in highly trained male rowers. *Hormone and Metabolic Research,* **35**, pp. 109-113.

Malina, R., 1973, Menarche in athletes; a synthesis and hypothesis. *Annals of Human Biology,* **10**, pp. 1-24.

Malina, R.M., 1983 Biological maturity status of young athletes. In: Brown EW, Branta CF, editors. *Competitive Sports for Children and Youth.* (Champaign, IL.: Human Kinetics), pp. 121-140.

Mansfield, M.J. and Emans, S.J., 1993, Growth in female gymnasts: should training
decrease during puberty? *Journal of Pediatrics,* **122**, pp. 237-240.

Mero, A., Jaakkola, L. and Komi, P.V., 1990, Serum hormones and physical performance capacity in young boy athletes during a 1-year training period. *European Journal of Applied Physiology,* **60**, pp. 32-37.

Okasora, K., Takaya, R., Tokuda, M., Fukunaga, Y., Oguni, T., Tanaka, H., *et al.,* 1999, Comparison of bioelectrical impedance analysis and dual energy X-ray absorptiometry for assessment of body composition in children. *Pediatrics International,* **41**, pp. 121-125.

Rich, P.A., Villani, R., Fulton, A., Ashton, J., Bass, S., Brinkert, R., *et al.,* 1992, Serum cortisol concentration and testosterone to cortisol ratio in elite prepubescent male gymnasts during training. *European Journal of Applied Physiology,* **65**, pp. 399-402.

Rilling, J.K., Worthman, C.M., Campbell, B.C., Stallings, J.F. and Mbizva, M., 1996, Ratios of plasma and salivary testosterone throughout puberty: production versus bioavailability. *Steroids,* **61**, pp. 374-378.

Schall, J.I., Semeao, E.J., Stallings, V.A. and Zemel, B.S., 2002, Self-assessment of
sexual maturity status in children with Crohn's disease. *Journal of Pediatrics,* **141**, pp. 223-229.

Sung, R.Y., Lau, P., Yu, C.W., Lam, P.K. and Nelson, E.A., 2001, Measurement of
body fat using leg to leg bioimpedance. *Archives of Disease in Childhood,* **85**, pp. 263-267.

Tanner, J.M., 1962, *Growth at adolescence.* (Oxford, UK: Blackwell Scientific Publications).

Theintz, G., Ladame, F., Howald, H., Weiss, U., Torresani, T. and Sizonenko, P.C.,
1994, L'enfant, la croissance et le sport de haut niveau. *Schweizerische Zeitschrift
fur Medizin und Traumatologie,* pp. 7-15.

Theintz, G.E., 1994, Endocrine adaptation to intensive physical training during growth. *Clinical Endocrinology,* **41**, pp. 267-272.

Theintz, G.E., Howald, H., Weiss, U. and Sizonenko, P.C., 1993, Evidence for a reduction of growth potential in adolescent female gymnasts. *Journal of Pediatrics,* **122**, pp. 306-313.

Tofler, I.R., Stryer, B.K., Micheli, L.J. and Herman, L.R., 1996, Physical and emotional problems of elite female gymnasts. *New England Journal of Medicine,* **355**, pp. 281-283.

Varona-Lopez, W., Guillemot, M., Spyckerelle, Y., Mulot, B. and Deschamps, J.P.,
1988, Self assessment of the stages of sex maturation in male adolescents [In French]. *Pediatrie,* **43**, pp. 245-249.

Warren, M.P., 1980, The effects of exercise on pubertal progression and reproductive function in girls. *Journal of Clinical Endocrinology and Metabolism,* **51**, pp. 1150-1157.

Wilson, J.D., George, F.W. and Griffin, J.E., 1981, The hormonal control of sexual development. *Science,* **211**, pp. 1278-1284.

Zakas, A., Mandroukas, K., Karamouzis, M. and Panagiotopoulou, G., 1994, Physical training, growth hormone and testosterone levels and blood pressure in prepubertal, pubertal and adolescent boys. *Scandinavian Journal of Medicine and Science in Sports,* **4**, pp. 113-118.

CHAPTER NINE

Body Composition Before and After Six Weeks Pre-season Training in Professional Football Players

E. Egan, T. Reilly, P. Chantler and J. Lawlor
Research Institute for Sport and Exercise Sciences, Liverpool John Moores University

1. INTRODUCTION

The relative amounts of lean and fat mass are important factors in games players preparing for competitive performance. Lean mass contributes to power production during high-intensity activity whilst extra fat mass constitutes added load to be lifted against gravity during jumping and locomotion. Recent technological advances have enhanced accuracy and precision in the measurement of body composition. The emergence of dual energy x-ray absorptiometry (DEXA) as a body composition tool has promoted much interest in the field of anthropometry due to its unique ability to subdivide the body into separate components of bone mineral mass, fat mass and fat-free mass. Additionally, it overcomes the population-specific nature of equations for predicting body fat from anthropometric measures, and the assumptions of constant fat-free tissue density associated with hydrodensitometry.

Despite its advantages, DEXA is not considered a suitable measurement tool for use in field conditions and its relative unavailability means that it is still not a popular technique among sport science support staff. The preference in field-work has been for indirect methods such as skinfold thicknesses (Reilly et al., 1996). As DEXA measures only masses, not volumes, the uncertainty therefore is not of its failure in measuring volume, but in assumptions necessary to relate DEXA tissue masses to volumetric measurements from other methods in multi-compartment models. The other shortcoming of DEXA is predicting soft tissue mass in pixels containing bone, via an assumed fat distribution model. The thorax has relatively few non-bone pixels, and predictions of fat on the torso are problematic. These are, however, specific to the DEXA manufacturer, scanner model and software version.

It has been reported that DEXA fails to identify lard placed on the trunk (Salamone et al., 2000), though Kohrt (1998) reported that DEXA correctly identified 96% of exogenously placed lard as fat irrespective of whether it was placed on the legs or torso. The aims of this study, therefore, were twofold: i) to determine changes in body fat, muscle mass and bone density measurements over 6 weeks of pre-season training in professional football players and ii) to compare body fat percentages derived from skinfold measurements and from DEXA whole-body measurements.

2. METHODS

2.1 Subjects

Participants were male professional English Premier League association football players (n = 16), including 3 goalkeepers, who underwent body composition assessment as part of routine pre-season assessment. The mean (±SD) characteristics were: age: 26.3 ± 5.0 years; mass: 85.0 ± 6.6 kg; height: 1.84 ± 0.05m. Written consent to participate was received from all players prior to testing.

2.2 Body Fat Estimation

Whole-body fat mass, percentage body fat, bone mineral density and fat-free soft tissue mass data were measured according to standard operating procedures using a fan beam dual energy x-ray absorptiometry scanner (DEXA; Hologic QDR series Delphi A, Bedford, Massachusetts). Scans were analysed using system Hologic QDR software for Windows version 11.2 (© 1986-2001 Hologic Inc.) according to standard analysis protocols. Regional fat mass and lean mass measurements were derived from whole-body scans, using standard analysis methods. Participants wore light clothing without zippers or metal buttons and removed all jewellery for the scanning procedures. Ethical approval for this technique was obtained from the University's Human Ethics Committee.

Table 1 Reliability measures for DEXA body composition assessment (n = 36)

Variable		CV (%)	TEM	TEM (%)
Fat mass	Whole-body	1.9	0.37 kg	1.6
	Left arm	2.8	0.06 kg	3.8
	Right arm	3.1	0.07 kg	4.4
	Trunk	1.9	0.42 kg	3.2
	Left leg	2.0	0.15 kg	3.1
	Right leg	2.4	0.18 kg	2.9
	Total arms	2.1	0.09 kg	3.6
	Total legs	1.8	0.27 kg	2.3
Fat-free soft tissue mass	Whole-body	1.0	0.44 kg	1.0
	Left arm	4.9	0.06 kg	2.8
	Right arm	4.5	0.05 kg	3.1
	Trunk	3.2	0.26 kg	1.9
	Left leg	3.0	0.11 kg	2.0
	Right leg	2.8	0.11 kg	2.4
	Total arms	3.6	0.85 kg	2.0
	Total legs	2.3	0.17 kg	1.8
Percent Fat	Whole-body	1.9	0.41 %	1.6
BMD	Whole-body	1.3	0.01 g·cm^{-2}	1.1
Mass	Whole-body	0.7	0.45 kg	0.7

Skinfold measurements were taken by a trained anthropometrist at four sites (biceps, triceps, subscapular and suprailiac spine). Percent body fat was estimated using the equations of Durnin and Womersley (1974).

$$CV = \frac{\left(\frac{SD_{diff}}{\sqrt{2}}\right)}{X} \times 100$$

where SD_{diff} is the standard deviation of the total differences between the measures; and $\overline{X}$ is the mean of all values

Equation 1 Coefficient of variation (CV)

The coefficients of variation (CV; Equation 1) of duplicate whole-body DEXA measurements in our laboratory ranged between 1% and 1.9% for bone mineral density, total mass, fat-free soft tissue mass, fat mass and percentage fat. Segment variation was less than 4.9% for fat mass and less than 3.1% for fat-free soft tissue mass. These values, together with the technical error of measurements (TEM; Equation 2), are tabulated in Table 1.

$$TEM = \sqrt{\frac{\sum d^2}{2n}}$$

where d is the difference between each pair of measures; and n is the sample number

Equation 2 Technical error of measurement (TEM)

2.3 Assessment Protocol

Duplicate DEXA measurements were carried out on 13 players at the beginning of the pre-season period of training (late June) and six weeks later at the end of the pre-season programme (early August). Skinfold measurements were taken on these and an additional 3 players, who also underwent DEXA assessment, at the second visit only. This schedule is outlined in Figure 1.

Late June   *Pre-season training (6 weeks)*   Early August

TEST 1	TEST 2
N = 13	N = 13 + 3
DEXA	DEXA
	Skinfolds

Figure 1 Schedule for body composition assessment

2.4 Statistics

Comparisons between body composition variables for first and second visits were made using t-tests for matched samples. Bias between the two methods of estimating percent body fat was examined using a Bland-Altman plot. A paired t-test and correlation analysis were also performed to assess the relationship between the two methods.

3. RESULTS

3.1 Changes in Body Composition Following Six Weeks Pre-season Training

Following six weeks pre-season training, percent body fat values, as measured by DEXA, were reduced in all players, the mean difference of 2.0% body fat being significant ($P < 0.001$). Actual body fat values were reduced by 2.0 kg, a 13.1% change from initial values. Whole-body bone mineral density values showed a significant increase over the pre-season training period ($P < 0.01$). Increases in lean mass were not significant ($P > 0.1$).

When analysed according to body segment, reduction in fat mass following the six weeks training was significant at all sites except the right leg. The mean percentage change in fat mass was 21.5%, 18.3% and 13.1% for arm, trunk and leg regions respectively. This decrease equates to a fat mass loss of 0.3 kg, 1.0 kg, 0.7 kg at the arms, trunk and legs respectively. Fat-free soft tissue mass did not change significantly with training at any of the segments.

Table 2 Anthropometric values (mean ± SD) before and after 6 weeks training (n = 13).

Variable	Pre-training	Post-training
Anthropometric Measures		
Mass (kg)**	86.6± 7.0	84.7 ± 6.9
Sum of skinfolds (mm)	-	30.2 ± 4.5
Body fat (%)	-	13.0 ± 2.0
DEXA Measures		
Body fat (%)**	15.3 ± 1.9	13.3 ± 1.6
Fat mass (kg)**	13.3 ± 2.1	11.3 ± 1.8
Fat-free soft tissue mass (kg)	69.6 ± 5.6	69.7 ± 5.7
Bone mineral density (kg)*	1.43 ± 0.1	1.46 ± 0.1

* denotes significance p < 0.01;
** denotes significance p < 0.001

3.2 Estimation of Body Fat by DEXA and Skinfold Measurements

A Bland-Altman plot of the two methods is shown in Figure 2

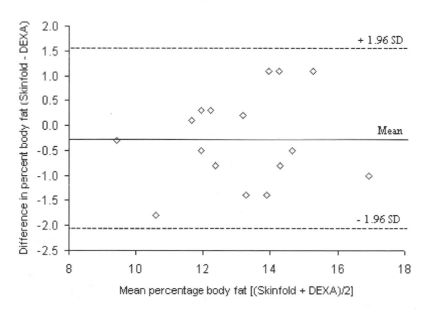

Figure 2 Bland-Altman plot showing mean values and 95% limits of agreement for DEXA and skinfold measures against the difference between the two methods. (n = 16)

A paired-samples t-test showed that body fat estimates did not differ significantly between the two methods of measurement (13.0 ± 2.0 % for skinfolds compared with 13.3 ± 1.8 % for DEXA; P > 0.05). This similarity was also true when the head was excluded from analysis (13.0 ± 2.0 % for skinfolds compared with 12.8 ± 2.0 % for DEXA subtotal; P > 0.05).

Body fat estimates from DEXA were significantly lower when the head was excluded, than when it was included (P < 0.001), although the decrease averaged only 0.2%. Body fat estimates for DEXA and skinfold methods were highly correlated both when the head was included (r = 0.89, P < 0.001) and excluded (r = 0.88, P < 0.001). Mean sum of skinfolds was 30.5 ± 5.1 mm.

4. DISCUSSION AND CONCLUSIONS

Dual energy x-ray absorptiometry appears sensitive enough to detect body fat changes over 6 weeks of training. Change can be detected if the confidence interval (68% for +/- 1standard error; 95% for +/- 2 standard errors, where 1 SE = TEM x √2) for the change in measurements does not include zero. Due to slightly poorer precision in regions, the confidence of detecting regional change is less than that of the whole body. The changes of the present study were relatively small compared with data obtained in previous decades. White et al. (1988) reported mean body fat values of 19.3% using the Durnin and Womersley (1974) equations in top league players, 3 weeks from the beginning of the season. Comparison with our results suggests that 20 years on, players are returning for the pre-season period in superior physical condition, having adopted a maintenance training programme in the off-season. It seems also that anthropometric measurements do not change substantially during the competitive season. Casajús (2001) conducted a fitness assessment of a top Spanish League (La Liga) team in September after 5 weeks training and repeated the tests on 15 players in the following February. During this period body fat estimated from anthropometry decreased from 8.6 ± 0.91 to 8.2 ± 0.91% (P < 0.01) lean body mass increased from 71.9 ± 6.01 to 72.1 ± 5.77 kg whilst there was no significant change in body mass. A comparison with the current data would raise a question about the effectiveness of the pre-season regimes used by the players we have studied in terms of lean body mass.

There were changes in total body mass during the pre-season period amounting to an average loss of 1.9 kg. These differences were accounted for by the increased bone mineral density and mostly by the reduction in body fat. Fat-free soft-tissue mass remained unaffected during the training period, reflecting the bias towards aerobic training to the possible neglect of strength training during the pre-season period.

Body fat changes were evident at all anatomical regions, though the degree of loss was relatively greater in the arms (21.5% fat mass loss), compared with the trunk (18.3% loss) and the legs (13.1% loss). There was no significant change in lean mass at any of the sub-regions. While the ability of DEXA to divide the body into sub-regions has the potential to provide a useful insight into regional fat distribution and changes over time, it should be noted that the reliability of sub-region measurements is less than for whole-body scans due to the introduction of human error in the positioning of divisional lines between segments.

The British Olympic Association (Reilly et al., 1996) has recommended using anthropometric skinfold measures in the absence of access to DEXA or hydrostatic weighing facilities. The high correlation between DEXA and skinfold estimates suggests that despite apprehensions about this method of estimating adiposity (Clarys et al., 1987), it does appear to be a suitable tool in the field setting. It must be stressed that all measures were carried out by a trained anthropometrist.

The correlation between DEXA and skinfold estimates of body fat was not different when the head was included or excluded from DEXA estimates. Due to the impermeability of the skull to x-rays, brain fat is estimated using the DEXA software so there is a question of whether the estimate for the head value should be included. While this estimation did not have any visible effect on the current results, other than it gave a slightly higher mean value, the estimation of brain fat in total-body scans may affect the results in a sample where there is greater inter-individual variation of body fat percentage.

The bone mineral density observed in the current sample was relatively high, being well in excess of equipment reference values. Association football is a contact sport, many incidents occurring during games when possession of the ball is vigorously contested. It would seem reasonable that skeletal strength would afford some protection against physical injury in competition. The impact nature of many of the game's activities, for example quick decelerations, jumping, kicking, and tackling, offer stimuli for bone growth. The positive effect of football training is denoted by the significant increase in bone mineral density observed during this study and supported by previous reports of increased BMD in prepubescent (Vicente-Rodriguez et al., 2003) and adult (Wittich et al., 1998; Calbet et al., 2001) players.

The high BMD values reported in these elite athletes may have implications for the estimation of body fat by hydrostatic weighing where non-fat tissues are presumed to be of constant density. This caveat is especially true when the equations utilised to calculate body fat percentage have not been derived from a similarly athletic population. The use of skinfold equations which have not been derived from similar populations to the subject group, should be treated with caution. While the equation of Durnin and Womersley (1974) utilised in this study had been recommended by the steering groups of the British Olympic Association (Reilly et al., 1996), it is noted that these equations were not derived from an athletic population and did not include any lower limb measurements. Large changes in composition of the lower body, which are not similar to upper body changes, may therefore not have been detected by serial skinfold measurements. Further investigation is required.

In summary, DEXA proved to be sufficiently sensitive to detect changes in body fat and bone mineral density over a relatively short period of time. As it extends the range of information available compared to alternative methods, DEXA can be recommended for serial assessments of body composition in association football and other professional games. Secondly, body fat estimated by means of skinfold thicknesses can yield high correlations with values derived by

DEXA when trained experts are employed. Skinfold assessments therefore appear to be valid for use in field settings.

REFERENCES

Calbet, J.A., Dorado, C., Diaz-Herrera, P. and Rodriguez-Rodriquez, L.P., 2001, High femoral bone mineral content and density in male football (soccer) players. *Medicine and Science in Sports and Exercise*, **33**, pp. 1682-1687.

Casajús, J.A., 2001, Seasonal variation in fitness variables in professional soccer players. *Journal of Sports Medicine and Physical Fitness*, **41**, pp. 463-467.

Clarys, J.P., Martin, A.D., Drinkwater, D.T. and Marfell-Jones, M.J., 1987, The skinfold: myth and reality. *Journal of Sports Sciences*, **5**, pp. 3-33.

Durnin J. and Womersley J., 1974, Body fat assessment from total body density and its estimation from skinfold thickness: measurements on 481 men and women aged 16 to 72 years. *British Journal of Nutrition*, **32**, pp. 77-97.

Kohrt, W.M., 1998, Preliminary evidence that DEXA provides an accurate assessment of body composition. *Journal of Applied Physiology*, **84**, 372-377.

Reilly, T., Maughan, R.J., Hardy, L., 1996, Body fat consensus statement of the steering groups of the British Olympic Association. *Sports Exercise and Injury*, **2**, pp. 46-49.

Salamone, L.M., Fuerst, T., Visser, M., Kern, M., Lang, T., Dockrell, M., Cauley, J.A., Nevitt, M., Tylavsky, F. and Lohman, T.G., 2000, Measurement of fat mass using DEXA: a validation study in elderly adults. *Journal of Applied Physiology*, **89**, pp. 345-352.

Vicente-Rodriguez, G., Jimenez-Ramirez, J., Ara, I., Serrano-Sanchez, S., Dorado, C. and Calbert, J.A.L., 2003, Enhanced bone mass and physical fitness in prepubescent footballers. *Bone*, **33**, pp. 853-859.

White, J.E., Emery, T.M., Kane, J.E., Groves, R. and Risman, A.B., 1988, Pre-season fitness profiles of professional soccer players. In: Science and Football, edited by: Reilly, T., Lees, A., Davids, K., Murphy, W.J., London E. & F.N. Spon, pp. 164-171.

Wittich, A., Mautalen, C.A., Oliveri, M. B., Bagur, A., Somoza, F. and Rotemberg, E., 1998, Professional football (soccer) players have a markedly greater skeletal mineral content, density and size than age- and BMI-matched controls. *Calcified Tissue International*, **63**, pp. 112-117.

CHAPTER TEN

Body image and body composition differences in Japanese and Australian males

Kagawa, Masaharu, Kerr, Deborah, Dhaliwal, Satvinder S., Binns, Colin W.

Curtin University of Technology, School of Public Health, Perth, Australia

1. INTRODUCTION

Poor body image has been linked with eating disorders and is included as one of the diagnostic criteria for eating disorders (American Psychiatric Association, 2000). Therefore, how accurately a person perceives his or her own body is important in identifying those at risk of eating disorders. Females are more dissatisfied with their body and have a greater desire to be thinner than males. The incidence of eating disorders is also higher in females than in males (Craig and Caterson, 1990, Paxton et al., 1991, Brenner and Cunningham, 1992). As a result, much of the research to date has focused on females. There is however, evidence that the incidence of eating disorders is increasing in males (Olivardia et al., 1995, Braun et al., 1999). Studies have indicated that young males are not satisfied with their physique. Unlike females, males either wish to lose weight or they want to be more muscular (Drewnowski and Yee, 1987, Paxton et al., 1991).

Males who have a distorted perception about themselves are at risk of developing a male-specific health problem known as "muscle dysmorphia" (Pope Jr et al., 1997). This is a condition in which the individual is overly preoccupied with being more muscular and will either exercise excessively or abuse anabolic steroids to gain muscle mass. O'Dea (1995) has suggested that males who want to be more muscular and bigger are those who are small in physique. This suggests that males of ethnic backgrounds where a small physique is most common may have a greater desire to be more muscular, influenced by western values on what is considered a desirable physique. However, there has been only one cross-cultural study which has assessed the relationship between body perception and body composition using Caucasians living in different countries (Pope Jr et al., 2000) and no reported study comparing the relationship possessed by Caucasians and Asians. One study reported that approximately 40% of Japanese males who were classified as having an "average" amount of body fat or BMI regarded themselves as being overweight (Urata et al., 2001). These results differ from western studies which showed males are likely to underestimate their body fat (Blokstra et al., 1999, Donath, 2000) and suggests that ethnic differences in body perception between Japanese and Australian Caucasian males may exist.

To be able to assess the relationship between physique and body image accurate assessment of body composition is required. However, many body image studies have used only simple indices such as body mass index (BMI) or body weight to assess body composition. The BMI is a measurement of heaviness and cannot detect differences in body composition between individuals (Garn *et al.*, 1986). The aim of the study was to determine if there was a difference in how Japanese and Australian Caucasian males perceive their own body in comparison to their actual body composition assessed using anthropometry.

2. METHODS

Three groups of subjects were included in the study;

- 68 Japanese males living in Australia (JA),
- 72 Australian Caucasian males (AA), and
- 84 Japanese males living in Japan (JJ)

Australian Caucasian (AA) and Japanese males living in Australia (JA) were recruited in Perth (Western Australia) and Japanese males living in Japan (JJ) were recruited in Himeji (Hyogo prefecture). Subjects born in Japan, had Japanese citizenship, spoke Japanese as their first language and had Japanese parents were considered to be "Japanese" and were included in the study. The criteria for "Australian Caucasian" included possession of Australian nationality, parents were "Caucasian" (ie., not of a mixed parentage), and those who recognized themselves as of "Caucasian" origin. The study was approved by the Human Research Ethics Committee of Curtin University of Technology. The purpose of the study was explained to all subjects and after completing a written informed consent, all subjects underwent body composition assessment and completed questionnaires at the same occasion.

The questionnaire contained demographic and lifestyle questions adopted from previously validated questionnaires of Japanese National Nutrition Survey, Australian National Nutrition Survey, and the Hawaii' Cancer Research Survey. The questionnaire also included questions on their current body weight and body fatness. For body weight, the question asked subjects to select their response from a choice of "light", "slightly light", "average", "slightly heavy", "heavy", and "I don't know". Similarly, the question on body fatness had categories of "very small amount", "small amount", "average", "slightly large amount", "large amount", and "I don't know".

Subjects also completed the Somatomorphic Matrix (SM) assessment. This is a computer-based program that was developed by Gruber *et al.* (1998). The program provides a range of illustrations of different physiques (as shown in Figure 1) and asks subjects to select the image that is most appropriate for the question given. An advantage of the program is that each illustration has values for percent body fat (%BF) and muscularity, using the fat free mass index (FFMI). The body composition values obtained from the SM program may not be comparable with the values obtained from the actual body composition assessments as no validation of the body composition images has been undertaken (Kagawa *et al.*, 2002). However, by not comparing the SM results with the measured values, the SM program was considered useful to compare body image of the different ethnic groups. In the current study, subjects were asked to select the image that most closely resembles their current physique (expressed as "current") and the ideal male physique which they thought would be most preferred by females (expressed as "ideal").

Figure 1. Examples of illustrations provided in the Somatomorphic Matrix (SM) computer program. Each illustration was given specific %BF and FFMI values.

Illustrations: A.J. Gruber and H.G. Pope, Biological Psychiatry Laboratory, McLean Hospital, Belmont MA, USA

Body composition was assessed using anthropometry according to the International Society for the Advancement of Kinanthropometry (ISAK) protocol (ISAK, 2001). Measurement sites included stature, body mass, eight skinfolds (triceps, subscapular, biceps, iliac crest, supraspinale, abdominal, front thigh, and medial calf), five girths (relaxed arm, flexed arm, waist, gluteal, and maximum calf), and four bone breadths (biacromial, biiliocristal, humerus, and femur). Japanese and Australian Caucasian males living in Australia were measured by a Level 3 anthropometrist accredited by ISAK and the measurements of Japanese males living in Japan were taken by three Level 1 and a Level 3 anthropometrists. All participating Level 1 anthropometrists demonstrated acceptable limits of technical error of measurements (Gore *et al.*, 1996) prior to their involvement in data collection. All anatomical landmarks were located and marked by the Level 3 anthropometrist (MK).

From the information obtained, the BMI, the waist to hip ratio (WHR), sum of skinfolds, and height-corrected sum of skinfolds were calculated. For calculation of %BF, the Durnin and Womersley body density prediction equation (Durnin and Womersley, 1974) was used to estimate their body density and this result was then converted into %BF using Siri's equation (Siri, 1961). Together with the body fat mass index (BFMI), the FFMI [fat free mass (kg)/stature $(m)^2$] was first proposed by VanItallie *et al.* (1990) as a useful tool in nutritional assessment. In the current study the FFMI was used as an indicator of muscularity as it was used in the SM program. The FFMI was calculated using the equation proposed by Kouri, *et al.* (1995) [body mass*$(100 - \%BF)/100 + 6.1*(1.8\text{-stature})/\text{stature}^2$] in order to make an assessment of the subject's muscularity consistent with the what is assessed with the SM program.

All analysis was conducted using SPSS statistical package for Windows (version 10.0, 1999, Chicago). One-way ANOVA was conducted to assess differences in body composition between study groups. To assess the accuracy of how the subjects perceived their body in relation to their measured body composition, weighted Kappa analysis was conducted for their perceived "heaviness" and "fatness" using the BMI as an index for "heaviness" and %BF and sum of skinfolds as indices for fatness respectively. Cut-off points for the BMI, %BF and sum of skinfolds are presented in Table 1.

Table 1. Cut-off points using the body mass index (BMI), percent body fat (%BF) and sum of skinfolds for the weighted kappa assessments, comparing Australian males (AA) with Japanese males living in both Australia (JA) and Japan (JJ).

Cut-off points	Light	Average	Heavy/Large
BMI (kg/m^2):			
JA	< 18.5	18.5 - 22.9	≥ 23.0
JJ	< 18.5	18.5 - 22.9	≥ 23.0
AA	< 18.5	18.5 - 24.9	≥ 25.0
Percent body fat (%):			
All groups	< 10	10 - 19.9	≥ 20
Skinfold sum (mm):[*]			
JA (n=145)	< 59.4	59.4 - 80.3	≥ 80.4
JJ (n=84)	< 62.4	62.4 - 80.3	≥ 80.4
AA (n=143)	< 66.3	66.3 - 97.7	≥ 97.8

* The numbers provided next to each study group are the number of subjects included in the database which used to determine the tertiles of skinfold cut-off values.

For the Australian Caucasian males the WHO BMI classification (WHO, 2004) was used and for the Japanese the BMI classification proposed specifically for the population of the Asia-Pacific region (the WPRO criteria) (WHO/IASO/IOTF, 2000). It has been suggested that Asians may have a greater %BF at the same BMI values compared to Caucasians (Deurenberg et al., 2002, Deurenberg et al., 2003, WHO, 2004). While the current study used the BMI as an indicator of heaviness, the BMI cut-off points that reflect %BF of each ethnic group were used in order to maintain the applicability of the BMI as an effective screening tool for obesity. Both 10% and 20% of %BF were proposed as the cut-off points for %BF based on the values previously suggested by both Japanese and Western researchers (Nagamine, 1972, Wilmore et al., 1986). Cut-off points for the sum of skinfolds were based on tertiles proposed from the dataset available (Kagawa, 2004).

3. RESULTS

Ethnic differences in body composition between the groups are shown in Table 2. Australian males were significantly ($p<0.05$) taller and heavier than Japanese males. Australian Caucasian males were also significantly greater in the FFMI value than their Japanese counterparts, indicating a greater muscularity of Australian males than Japanese. The Australian males had higher mean BMIs and sums of skinfolds, than the Japanese groups. The %BF of the Australians and the Japanese males living in Australia were similar, but the results of both groups were significantly higher than the %BF of Japanese males living in Japan. While there were no significant differences in %BF between the JA and the AA groups, the AA group showed greater sum of skinfolds even after adjustment for stature.

Table 2. Results of body composition assessment comparing Australian males (AA) with Japanese males living in both Australia (JA) and Japan (JJ).

Subject characteristic	JA Mean ± SD	JJ Mean ± SD	AA Mean ± SD
Number of subjects	68	84	72
Age (Years)	23.5 ± 2.9	20.5 ± 1.7[#, ^]	23.1 ± 3.3
Stature (cm)	171.7 ± 5.1	172.9 ± 5.4	182.2 ± 7.2[*, #]
Body mass (kg)	64.2 ± 7.6	64.0 ± 9.1	80.6 ± 11.9[*, #]
Body Mass Index (kg/m^2)	21.8 ± 2.3	21.4 ± 2.8	24.3 ± 3.1[*, #]
Waist Hip Ratio (WHR)	0.81 ± 0.04	0.79 ± 0.04[#, ^]	0.82 ± 0.04
Total Body Fat (%BF)	17.0 ± 5.1	16.5 ± 5.1[#]	19.1 ± 6.0
Sum of 8 skinfolds (mm)	81.8 ± 33.6	79.6 ± 37.9	106.5 ± 47.1[*, #]
Ht-corrected sum of 8 skinfolds (mm)	81.2 ± 33.8	78.3 ± 36.7	99.4 ± 44.0[*, #]
Fat Free Mass Index (kg/m^2)	18.2 ± 1.3	17.9 ± 1.6	19.4 ± 1.6[*, #]

* Significant difference between AA and JA at the 0.05 level.
# Significant difference between AA and JJ at the 0.05 level.
^ Significant difference between JA and JJ at the 0.05 level.

The results of perceived "current" and "ideal" %BF and FFMI values using the Somatomorphic Matrix (SM) computer program are shown in Table 3. The results indicate that males desire a more muscular physique regardless of their ethnic backgrounds (Statistically significant for the JJ and JA groups).

Table 3. Fatness[1] and muscularity[2] values obtained from perceived "current" and "ideal" physiques using the Somatomorphic Matrix (SM) computer program. Results are mean ± SD.

	JA Mean ± SD	JJ Mean ± SD	AA Mean ± SD
Body fat (%):			
Current physique	19.6 ± 7.9[*]	17.7 ± 9.1	18.5 ± 8.5[*]
Ideal physique	17.1 ± 5.6	16.4 ± 4.8	15.3 ± 6.2
FFMI (kg/m^2):			
Current physique	20.7 ± 2.6[*]	20.7 ± 2.4[*]	20.7 ± 2.5[*]
Ideal physique	23.2 ± 2.5	22.7 ± 1.9	22.8 ± 2.3
Differences (ideal-current):			
%BF difference	-2.5 ± 9.2	-1.3 ± 10.0	-3.2 ± 9.8
FFMI difference	2.4 ± 3.2	2.0 ± 2.4	2.1 ± 2.8

* Significant differences between "current" and "ideal" %BF and FFMI values at the 0.05 levels.

[1] Percent body fat (%BF) was used as an indicator for fatness.

[2] Fat Free Mass Index (FFMI) was used as an indicator for muscularity.

To assess how subjects perceived their own body in relationship to their actual body composition, the following pairs were examined using a weighted Kappa to assess the level of agreement of the following (Figure 2): (1) Perceived heaviness to the BMI; (2) Perceived fatness to %BF; (3) Perceived fatness to sum of skinfolds. The results indicate that all groups had relative understanding between their perceived heaviness and their actual body mass (agreement ranged between 0.42-0.55). However, none of the groups showed good agreement between their perception of own fatness and their actual %BF (agreement ranged between 0.28-0.35). The JJ group showed a moderate level of agreement between their perceived fatness and sum of skinfolds (0.51) that was similar to the agreement obtained between a perceived heaviness and the BMI.

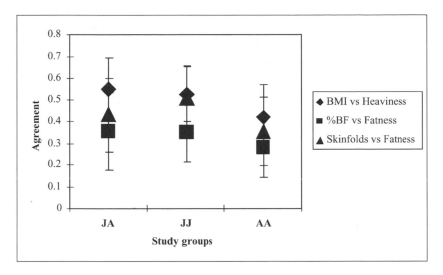

Figure 2. Differences in weighted Kappa results with 95% Confidence limits between perceived heaviness and fatness in relation to the BMI and %BF comparing Australian males (AA) with Japanese males living in both Australia (JA) and Japan (JJ).

Table 4 illustrates how subjects perceived how heavy they were in relation to their BMI values. Approximately 15% of Japanese males who were classified as "average" (BMI 18.5-22.9 kg/m^2), perceived themselves as being "heavy", compared to only 6.7% of Australians in the same category (BMI 18.5-24.9 kg/m^2). Similarly, approximately 20% of JA and JJ subjects who were classified as "average" using the %BF classification perceived themselves as being fat, compared to only 4.8% of Australian Caucasian males in the same category (Table 5).

When the sum of skinfolds was used as an indicator for fatness, approximately 50% of Japanese males whose sum of skinfolds was considered "average" perceived themselves correctly (Table 6). However, more than 25% of the JA and the JJ groups who were classified as "average" perceived themselves as being fat. In addition, 50% of the JA subjects whose sum of skinfolds was in the lower tertile perceived themselves as having an average amount of body fat. In contrast to the results of the JA group, the JJ individuals in the lower tertile of sum of skinfolds did not show the same pattern. This suggests that the JA subjects have a greater distortion of how they perceive their body than the JJ subjects.

More than 30% of Australian Caucasian males who were classified as overweight based on the BMI classification perceived themselves as not heavy. Similarly, more than 50% of those classified as overweight using the %BF classification perceived themselves as not fat. While most of the JA (89.5%) and JJ (100%) subjects in the overweight category perceived themselves as heavy, only 65.4% of the AA subjects perceived themselves correctly (Table 4). Similarly, a smaller proportion of the Australian males compared to the Japanese males perceived themselves as being fat, even though they were classified as overweight using the sum of skinfolds or %BF classifications. These results indicate that Australian Caucasian males are more likely to underestimate their body fat whereas Japanese males are more likely to overestimate their body fat.

Table 4. Proportion of subjects with correct body perception, comparing Australian males (AA) with Japanese males living in both Australia (JA) and Japan (JJ): Perceived heaviness using the BMI as an indicator of heaviness.

JA		Perceived heaviness (%)		
		Light	Average	Heavy
Actual	Light	100.0	0.0	0.0
BMI	Average	32.6	53.5	14.0
category	Heavy	0.0	10.5	89.5
JJ		Perceived heaviness (%)		
		Light	Average	Heavy
Actual	Light	100.0	0.0	0.0
BMI	Average	41.1	42.9	16.1
category	Heavy	0.0	0.0	100.0
AA		Perceived heaviness (%)		
		Light	Average	Heavy
Actual	Light	100.0	0.0	0.0
BMI	Average	33.3	60.0	6.7
category	Heavy	7.7	26.9	65.4

Note: For the Japanese males, the WPRO criteria, which was proposed specifically for the population living in the Asia-Pacific region, was used with the cut-off points of <18.5, 18.5-22.9, and ≥23 for categories of "underweight", "average", and "overweight". For Australian Caucasian males the WHO criteria was used with the cut-off points of <18.5, 18.5-24.9, and ≥25 for categories of "underweight", "average", and "overweight".

Table 5. Proportion of subjects with correct body perception, comparing Australian males (AA) with Japanese males living in both Australia (JA) and Japan (JJ): Perceived fatness using %BF calculated from anthropometry.

JA		Perceived fatness (%)		
		Less fat	Average	Fat
Actual %BF	Less fat	66.7	0.0	33.3
category	Average	25.0	54.5	20.5
	Fat	11.1	16.7	72.2
JJ		Perceived fatness (%)		
		Light	Average	Heavy
Actual %BF	Light	100.0	0.0	0.0
category	Average	40.0	41.7	18.3
	Heavy	0.0	22.2	77.8
AA		Perceived fatness (%)		
		Light	Average	Heavy
Actual %BF	Light	100.0	0.0	0.0
category	Average	40.5	54.8	4.8
	Heavy	19.2	38.5	42.3

Table 6. Proportion of subjects with correct body perception, comparing Australian males (AA) with Japanese males living in both Australia (JA) and Japan (JJ): Perceived fatness using the sum of skinfolds as an indicator of fatness.

JA		Perceived fatness (%)		
		Less fat	Average	Fat
Actual %BF	Less fat	45.0	50.0	5.0
category	Average	21.1	52.6	26.3
	Fat	30.4	65.2	4.4
JJ		**Perceived fatness (%)**		
		Light	Average	Heavy
Actual %BF	Light	66.7	33.3	0.0
category	Average	21.4	50.0	28.6
	Heavy	11.5	23.1	65.4
AA		**Perceived fatness (%)**		
		Light	Average	Heavy
Actual %BF	Light	66.7	33.3	0.0
category	Average	30.4	65.2	4.4
	Heavy	21.9	40.6	37.5

4. DISCUSSION

This study was the first to compare the body image of Japanese and Australian males in relation to their actual body composition as determined by anthropometry. Australian males are more likely to underestimate their body fat whereas Japanese males are more likely to overestimate their body fat. However, both Japanese and Australian Caucasian males selected a more muscular and leaner physique as their ideal which differed from their perception of their current physique. The findings were consistent with previous studies that suggested males' desire to be muscular (Silberstein *et al.*, 1988, Paxton *et al.*, 1991). These studies however, failed to quantify the subjects' actual body composition and how it related to their desired physique. By using more comprehensive anthropometry to assess body composition, this current study confirmed that males desire to become bigger by increasing muscle mass.

Ethnicity and a country of residence were found to have little effect on what males perceived as their ideal physique. This may indicate that their desire to become more muscular is associated with a gender instinct. The increasing influence of Western culture on Japanese males may also be affecting their perceived ideal physique. Because however, there are no historical data on the body image of Japanese males it is difficult to verify the influence of Westernisation, particularly in the post second world war period.

Anthropometry was used to determine ethnic and environmental influences on how subjects perceived their level of body fat and how heavy they were in relation to their actual body composition. The agreement on heaviness between their perception and the BMI ranged from 0.42 to 0.55 across the study groups. Lower levels (0.28-0.35) of agreement were obtained for their perceived fatness in relation to their measured %BF. This indicates both Japanese and Australian males possess a relatively good understanding of how heavy they were but were less accurate in their assessment of the amount of body fat they had. This is not unexpected as males are more likely to weigh themselves frequently, but would be unlikely to have their body fat measured. Specific BMI classifications were used in this study, in order to control for ethnic differences in the BMI-%BF relationship. The finding of lower agreement for perceived fatness compared to their perceived heaviness also suggests that understanding how heavy they were was not necessarily linked to an understanding of their body fat levels.

In addition to %BF values estimated from anthropometry, a sum of eight skinfolds was also used as an indicator of fatness. In all groups, agreement between their perceived fatness and sum of skinfolds was higher than the agreement values with perceived fatness using %BF. The higher agreements between perceived fatness and sum of skinfolds compared to the results using %BF suggests that males can more accurately perceive their body fat using the sum of skinfolds. This may also indicate the error involved in estimating body fat from skinfold equations (Martin *et al.*, 1985, Norton and Olds, 1996). The only way to determine how closely %BF and skinfold sum are associated is to compare against %BF obtained from underwater weighing, dual energy x-ray absorptiometry or a multi-compartment model using both. These data were not available for this sample.

In comparison to Japanese males, Australian Caucasian males were less concerned about their current body mass or fatness. As a result, more than 30% of those with the BMI above 25 and more than 50% of those with %BF above 20% perceived themselves as "not heavy" or "not fat". These results are consistent with previous studies conducted in Western countries suggesting males underestimate their level of fatness (Valtolina, 1998, Blokstra *et al.*, 1999, Donath, 2000). In contrast, Japanese males did not underestimate their body fat to same degree, as seen in Australian males. Japanese males, regardless of their country of residence, expressed concern about their body mass and fatness. Approximately 15% of Japanese males whose BMI is within an acceptable level perceived themselves as being heavy (compared with 6.7% for the AA group) and 20% of those whose %BF was within an average level perceived themselves as fat (compared with 5% for the AA group). This overestimation of their own level of fatness among Japanese males is consistent with the results obtained by Urata *et al.* (2001). Differences in the proportion of individuals who overestimated their %BF may be associated with differences in sample size between studies as well as differences in methodology, including cut-off points of the BMI and body composition assessment methods.

The current study found ethnic differences in how males perceive their body in relation to their body composition. In addition, Japanese males living in Australia were more likely to overestimate their body fatness than those living in Japan. This may be due to the influence of a more western living environment. Rand and Kuldau (1990) suggested that body perception varies depending on ethnic and cultural variables. These differences in body perception may influence their awareness on their fatness and weight-management behaviours long-term.

The current results suggest a minimum influence of ethnicity on the selection of the ideal physique in males. This ideal physique selected by males may not be the same male physique which females perceive as ideal (Fallon and Rozin, 1985, Rozin and Fallon, 1988). It has previously been reported that Japanese females prefer males with "unisexual physique" (Takeda *et al.*, 1996). To explore gender differences in body image and body perception, future studies should include both male and female subjects. In addition, it is recommended that more comprehensive body composition assessments should be undertaken using skeletal muscle mass (Lee *et al.*, 2000) or dual energy x-ray absorptiometry. The current study used cut-off points for sum of skinfolds based on a relatively small sample of Japanese and Australian Caucasian males that were available. Therefore a larger sample of both genders and wider age group and body composition is needed.

5. CONCLUSION

The findings of this study indicate that both Japanese and Australian males possess a similar ideal male physique in relation to their perceived current physique. However, there were possible ethnic differences in the way Japanese and Australian Caucasian males perceived themselves in relation to their actual body composition measured by anthropometry. While Australian males underestimated their level of overweight, a relatively large proportion of Japanese males perceived themselves as being overweight even though they were not. No subjects were able to accurately assess their level of body fat, regardless of their ethnic background. Australian males were more likely to underestimate their body fat whereas Japanese males were more likely to overestimate their body fat. Further research is needed to identify gender and ethnic differences in body perception and to assess the impact of Westernisation on body image.

Acknowledgements

The authors express their gratitude to Associate Professor M. Ouchi (School of Engineering), Associate Professor H. Uchida, Miss T. Masuda and Miss T. Nishiwaki (School of Human Science and Environment) of University of Hyogo who assisted subject recruitment and data collection in Japan.

REFERENCES

American Psychiatric Association (Eds.), 2000, *Diagnostic and statistical manual of mental disorders-Fourth edition-Text revision.* (Washington, DC: American Psychiatric Association).

Blokstra A., Burns C.M. and Seidell J.C., 1999, Perception of weight status and dieting behavior in Dutch men and women. *The International Journal of Eating Disorders*, 23, pp. 7-17.

Braun D.L., Sunday S.R., Huang A. and Halmi K.A., 1999, More males seek treatment for eating disorders. *The International Journal of Eating Disorders*, 25, pp. 415-424.

Brenner J.B. and Cunningham J.G., 1992, Gender differences in eating attitudes, body concept, and self-esteem among models. *Sex Roles*, 27, pp. 413-437.

Craig P.L. and Caterson I.D., 1990, Weight and perceptions of body image in women and men in a Sydney sample. *Community Health Studies*, 14, pp. 373-381.

Deurenberg P., Bhaskaran K. and Lian P.L., 2003, Singaporean Chinese adolescents have more subcutaneous adipose tissue than Dutch Caucasians of the same age and body mass index. *Asia Pacific Journal of Clinical Nutrition*, 12, pp. 261-265.

Deurenberg P., Deurenberg-Yap M. and Guricci S., 2002, Asians are different from Caucasians and from each other in their body mass index/body fat percent relationship. *Obesity Reviews*, 3, pp. 141-146.

Donath S.M., 2000, Who's overweight? Comparison of the medical definition and community views. *The Medical Journal of Australia*, 172, pp. 375-377.

Drewnowski A. and Yee D.K., 1987, Men and body image: Are males satisfied with their body weight? *Psychosomatic Medicine*, 49, pp. 626-634.

Durnin J.V.G.A. and Womersley J., 1974, Body fat assessed from total body density and its estimation from skinfold thickness: measurements on 481 men and women aged from 16 to 72 years. *British Journal of Nutrition*, 32, pp. 77-97.

Fallon A.E. and Rozin P., 1985, Short report-Sex differences in perceptions of desirable body shape. *Journal of Abnormal Psychology*, 94, pp. 102-105.

Garn S.M., Leonard W.R. and Hawthorne V.M., 1986, Three limitations of the body mass index. *The American Journal of Clinical Nutrition*, 44, pp. 996-997.

Gore C., Norton K., Olds T., Whittingham N., Birchall K., Clough M., Dickerson B. and Downie L., 1996, Accreditation in anthropometry: an Australian model. In *Anthropometrica*, edited by K. Norton and T. Olds. (Sydney: University of New South Wales Press), pp. 395-411.

Gruber A.J., Pope Jr. H.G. and Borowiecki III J.J., 1998, The development of the somatomorphic matrix: A bi-axial instrument for measuring body image in men and women. In *The Sixth Scientific Conference of the International Society for the Advancement of Kinanthropometry*, edited by K. Norton, T. Olds and J. Dollman, (Adelaide: International Society for the Advancement of Kinanthropometry), pp. 217-231.

ISAK, 2001, *International standards for anthropometric assessment*. (Canberra: ISAK).

Kagawa M., 2004, *Ethnic and cultural influences on body composition, lifestyle and body image among males*. Unpublished Doctoral thesis. School of Public Health. (Perth: Curtin University of Technology).

Kagawa M., Kerr D. and Binns C., 2002, Validation of the Somatomorphic Matrix computer program for predicting body composition in Japanese and Australian Caucasian males. In *2002 Australian Conference of Science and Medicine in Sport*, edited by SMA, (Melbourne: Sports Medicine Australia), p. 57.

Kouri E.M., Pope Jr H.G., Katz D.L. and Oliva P., 1995, Fat-free mass index in users and nonusers of anabolic-androgenic steroids. *Clinical Journal of Sport Medicine*, 5, pp. 223-228.

Lee R.C., Wang Z., Heo M., Ross R., Janssen I. and Heymsfield S.B., 2000, Total-body skeletal muscle mass: development and cross-validation of anthropometric prediction models. *American Journal of Clinical Nutrition*, 72, pp. 796-803.

Martin A.D., Ross W.D., Drinkwater D.T. and Clarys J.P., 1985, Prediction of body fat by skinfold caliper: Assumptions and cadaver evidence. *International Journal of Obesity*, 9, pp. 31-39.

Nagamine S., 1972, Hikashibou kara no himan no hantei [Jap]. *Nihon Ishikai Zasshi*, 68, pp. 919-924.

Norton K. and Olds T. (Eds.), 1996, *Anthropometrica*. (Sydney: University of New South Wales Press).

O'Dea J.A., 1995, Body image and nutritional status among adolescents and adults-A review of the literature. *Australian Journal of Nutrition and Dietetics*, 52, pp. 56-67.

Olivardia R., Pope Jr H.G., Mangweth B. and Hudson J.I., 1995, Eating disorders in college men. *The American Journal of Psychiatry*, 152, pp. 1279-1285.

Paxton S.J., Wertheim E.H., Gibbons K., Szmukler G.I., Hillier L. and Petrovich J.L., 1991, Body image satisfaction, dieting beliefs, and weight loss behaviors in adolescent girls and boys. *Journal of Youth and Adolescence*, 20, pp. 361-379.

Pope Jr H.G., Gruber A.J., Mangweth B., Bureau B., de Col C., Jouvent R. and Hudson J.I., 2000, Body image perception among men in three countries. *The American Journal of Psychiatry*, 157, pp. 1297-1301.

Pope Jr H.G., Gruber A.M., Choi P., Olivardia R. and Phillips K.A., 1997, Muscle dysmorphia – An underrecognized form of body dysmorphic disorder. *Psychosomatics*, 38, pp. 548-557.

Rand C.S.W. and Kuldau J.M., 1990, The epidemiology of obesity and self-defined weight problem in the general population: Gender, race, age, and social class. *The International Journal of Eating Disorders*, 9, pp. 329-343.

Rozin P. and Fallon A., 1988, Body image, attitudes to weight, and misperceptions of figure preferences of the opposite sex: A comparison of men and women in two generations. *Journal of Abnormal Psychology*, 97, pp. 342 345.

Silberstein L.R., Striegel-Moore R.H., Timko C. and Rodin J., 1988, Behavioral and psychological implications of body dissatisfaction: Do men and women differ? *Sex Roles*, 19, pp. 219-232.

Siri W.E., 1961, Body volume measurement by gas dilution. In *Techniques for measuring body composition*, edited by J. Brozek and A. Henschel. (Washington, DC: National Academy of Sciences), pp. 108-117.

Takeda A., Suzuki K. and Muramatsu T., 1996, Bulimia symptoms in male college students compared with female students [Jap]. *Rinshou Seishin Igaku*, 25, pp. 1083-1089.

Urata H., Fukuyama Y. and Tahara Y., 2001, Physique and its recognition in male students. *Japanese Journal of School Health*, 43, pp. 275-284.

Valtolina G.G., 1998, Body-size estimation by obese subjects. *Perceptual and Motor Skills*, 86, pp. 1363-1374.

VanItallie T.B., Yang M.U., Heymsfield S.B., Funk R.C. and Boileau R.A., 1990, Height-normalized indices of the body's fat-free mass and fat mass: Potentially useful indicators of nutritional status. *American Journal of Clinical Nutrition*, 52, pp. 953-959.

WHO, 2004, Appropriate body-mass index for Asian populations and its implications for policy and intervention strategies. *Lancet*, 363, pp. 157-163.

WHO/IASO/IOTF, 2000, *Asia-Pacific perspective: Redefining obesity and its treatment.* Health Communications Australia Pty Ltd.

Wilmore J.H., Buskirk E.R., DiGirolamo M. and Lohman T.G., 1986, Body composition-a round table. *The Physician and Sports Medicine*, 14, pp. 144-162.

CHAPTER ELEVEN

The observed and perceived body image of female Comrades Marathon athletes.

Beukes, N.M.[1], van Niekerk, R.L.[1] & Lombard, A.J.J.[2]
[1]Department of Psychology, University of Johannesburg, South Africa
[2]Department of Sport and Human Movement Studies, University of Johannesburg, South Africa .

1. INTRODUCTION

This study examined the relationship between the observed (what their actual size was) and perceived body image (what they think about their bodies) of non-elite female Comrades Marathon athletes. These athletes had finishing times in excess of 9 hours in the Comrades Marathon which is an 11-hour race over approximately 89 kilometers. The ladies race was won in 2005 in a time of 5:58:50.

Body image refers to the manner in which people view their bodies and consists of the mental representations they have of them (Cash, 1996). These mental representations encompass perceptions, cognitions, affects and behavior related to appearance, and how they conform to the cultural ideal of beauty and attractiveness (Cash & Hicks, 1990). Body image forms an integral part of a person's body esteem and overall self-worth, and can be regarded as a loose mental representation of the body that does not consistently reflect the way the external world views a person (Cash, 1996). When athletes compare their own bodies and body images with those of others, they could experience a discrepancy between their perceived body image and the actual body characteristics optimally suited for their sport.

Female athletes who develop a heightened concern about their body weight and shape, as well as societal pressures from coaches and peers to achieve an ideal body weight and shape, are affected in particular (Drewnowski & Lee, 1987). Such concern and pressure could lead to health-risk behavior like eating disorders (Brenner & Cunningham, 1992), unhealthy dieting (Cash, 1994a), and abnormal eating and weight-control behavior (Davis & Fox, 1993). Studies by Garner and Rosen (1991) have confirmed the high risk of weight concern and eating disorders in sports such as marathon running. Clark, Nelson and Evans (1988) indicated that 13% of elite female runners reported a history of anorexia nervosa, while 34% experienced disturbed eating patterns. This is especially evident for women who participate in sports like marathon running, that require weight restriction as an important part of performance demands (Soap & Murphy, 1995).

Garner, Rosen and Barry (1998) showed that there is a small subgroup of runners who engage in pathologic weight control behaviors and have eating disorders. In this regard, female athletes suffering from body-image disturbance often put emphasis on a lean body and through its endorsement of excessive exercise, the sport environment may make it easier for them to develop health risk behavior, but more difficult to identify and treat them (Hollilman, 1991). Weight and Noakes (1987) related the risk of female marathon runners to develop eating disorders to the need to be thin for the sport. This need is encouraged by a perception that the thinner you are the faster and better you will be able to run. Parker, Lambert and Burlingame (1994), showed that female marathon runners, who engaged in pathologic weight-control behavior, had elevated scores on the Drive for Thinness subscale of the Eating Disorder Inventory (EDI). Although many runners and their coaches are convinced of this relationship, the incidence of pathologic weight-control in such sports could still be dangerous, since these athletes often have less than 10 percent body fat (Humphries & Gruber, 1986; Rosen, McKeag, Hough & Curley, 1986; Harris & Greco, 1990).

Zulu (1999) argued that South African female athletes have to contend with a stereotypical cultural attitude that a woman's place is in the home. Such an attitude towards women has a handicapping effect on these athletes, but could be overcome with support and understanding. Regardless of such an attitude, women's running is rapidly growing in South Africa. Since the inception in 1979 of a 'ladies only' race in Durban, the number of participants has increased steadily to 3,200 in 1992 (De La Rey & Paruk, 1993). In spite of this growth, very little is known about the effects of perceived and observed body image of women long distance runners in South Africa.

The purpose of this study therefore was an attempt to contribute to the literature, by investigating whether a difference exists between the perceived and observed body image in female Comrades Marathon athletes. The Comrades Marathon is an ultra long distance marathon, run annually over approximately 89 kilometers between the cities of Durban and Pietermartzburg in Kwazulu-Natal, South Africa. It is more than twice the normal marathon distance and has a challenging elevation change profile that is run up the one year and down the other. These unique features differ much from standard-length marathons and thus provide an opportunity to investigate the difference between perceived and observed body image of female runners who exercise for and participate in events that challenge their bodies to such an extent.

2. RESEARCH METHODOLOGY

Body image can be measured on both a perceived and observed level. Anthropometry (on the observed level), in conjunction with psychometric evaluation (on the perceived level) could be a useful objective tool for quantifying body size, shape and composition (Wells, 1991). The congruence or incongruence of these representations in a person's experience of gender and cultural context is judged in relation to a number of dynamic set points and trigger personal and interpersonal behavior (Cash & Pruzinsky, 1990) as representations of a body image.

Seventy five regular, non-elite female Comrades Marathon athletes from 25 athletic clubs in Johannesburg, South Africa, were approached in collaboration with the chairpersons of these clubs. Only 49 of the selected athletes were willing to participate in the study.

After signing a consent form, the subjects completed the Multidimensional Body-Self Relations Questionnaire (MBSRQ) (Cash, 1994b; Cash & Pruzinsky, 1990). Percentage body fat was assessed by a qualified sport scientist who measured the six skinfolds (subscapular, triceps, supraspinale, abdominal, front thigh and medial calf) according to the Montreal Olympic Games Anthropometrical Project (MOGAP) (Carter,1982) using a Harpenden Skinfold Caliper (ModelHSK-BI, British Indicators, England). This particular tool was selected on the basis that it was developed for athletes and provides for female athletes by applying the following formula: Fat%= ($\sum$6 skinfolds x 0.1548) + 3.58. Body Mass Index (BMI) was calculated as mass in kg divided by height in metres squared. This index is widely used and reported in large epidemiological studies and moderately correlates with body fat ($r=0.80$) (Cash, 1985, Brownell, 1982). Body weight was measured on a Seca Bella 840 scale (Seca Ltd, Birmingham, England) to the nearest 0.1 kilogram and standing height to the nearest 0.1 centimeter using a stadiometer.

2.1 The Multidimensional Body-Self Relations Questionnaire (MBSRQ)

The MBSRQ (Cash, Winstead & Janda, 1986) was chosen for the purposes of this study, as it was seen to be the most comprehensive and psychometrically-studied cognitive assessment of body image (Cash, 1994b). It is a 69-item self-report inventory for the assessment of self-attitudinal aspects of body image. Body image is conceived as an individual attitudinal disposition toward the physical self. The MBSRQ's Factor subscales reflect two dispositional dimensions – 'evaluation' and 'cognitive-behavioral orientation' – vis-à-vis each of the three somatic domains of 'appearance', 'fitness' and 'health/illness'. These subscales include:

- Health evaluation, indicating perceptions of physical health and/or the freedom from physical illness.
- Health orientation, indicating the amount of time spent trying to lead a healthy lifestyle.
- Illness orientation, which indicates the extent of reactivity to being and becoming ill.
- Appearance evaluation, which evaluates feelings of physical attractiveness or unattractiveness and satisfaction or dissatisfaction with a person's appearance.
- Appearance orientation, assesses the amount of time spent by a person to improve appearance.
- Fitness evaluation, indicates feelings of being physically fit or unfit.
- Fitness orientation, assesses the amount of time spent trying to be physically fit or athletically competent.

It also includes three special multi-item subscales, namely:
(a) The Body Areas Satisfaction Scale (BASS), measuring satisfaction-dissatisfaction with body areas and attributes, (b) The Overweight Preoccupation Scale, measuring fat anxiety, weight vigilance, dieting and eating restraint, and (c) The self-Classified Weight Scale, assessing self-appraisals of being 'very underweight' to being 'very overweight'. The reliability of the subscales ranges from 0.83 to 0.92 and possess acceptable internal consistency and stability (Cash,

1990). Forty response variables were used in this study (2 dispositional dimensions X (7+3) subscales on the MBSRQ X 2 somatic domains).

3. ANALYSIS

As a first step in this research, variables, including the BMI and subscales of the MBSRQ, were assessed for correlation using Pearson Correlations. Significant (2-tailed) analyses were performed to establish statistical significance at the 0.05 and 0.01 level. Because of the small sample size (n=49) and a larger number of correlations (n=55), subsequent corrections were made through Sequential Bonferonni adjustments to ensure significance at the 0.05 level.

As a second step, two categories of athletes were identified as underweight (n=26) and normal or overweight (n=23), according to fat percentage (See Table 3). These two groups were compared to find a correlation between their perceived and observed body image. Ten two-sided independent sample t-tests (see Table 4) were conducted on the two groups. To ensure an overall Type 1 error rate of 0.05, the level of significance was set at p = 0.05/10 = 0.005.

4. RESULTS AND DISCUSSION

The relevant descriptive and body composition characteristics for the subjects are represented in Table 1.

	Minimum	Maximum	Mean	Standard deviation
Age	25	55	38.4	6.09
Body mass (kg)	46	79	61.28	6.61
Height (cm)	149	174	164.85	5.79
Fat percentage	12.70	28.50	19.77	3.89
Body Mass Index	17.81	27.02	22.55	2.11

Table 1: Descriptive and Body Composition characteristics of the subjects.

It was decided, for the purposes of this study, that the results from the MBSRQ would be an indication of the participants' perceptions of their body image (what they think about their bodies). The anthropometric measurements would be used as an indication of the participants' observed body image and calculated as a BMI (what the actual size of the body is).

The correlations between the subscales of the MBSRQ and BMI are represented in Table 2. Body Mass Index (BMI) showed a positive correlation (0.707) with self-classified weight, indicating that the participants tend to perceive and label their own weight close in relation to that of their BMI.

A positive correlation was found between appearance evaluation and body areas satisfaction (0.73). Those participants who were happy with their physical appearance were generally happy with the size and various areas of their bodies. Participants who were unhappy with their physical appearance were dissatisfied with various body areas. Participants who felt unhappy about their appearance, tended to pay more attention to their appearance and tended to engage in more extensive grooming behaviour.

		BMI	AE	AO	FE	FO	HE	HO	IO	BAS	OP
Appearance evaluation (AE)	r	-.29*	1								
	Sig. (2-tail)	.041									
	Bonferonni	1.476									
Appearance orientation (AO)	r	-.23	.01	1							
	Sig. (2-tail)	.107	.275								
	Bonferonni	2.889	4.675								
Fitness evaluation (FE)	r	.01	.33*	-.18	1						
	Sig. (2-tail)	.923	.019	.204							
	Bonferonni	0.923	0.760	3.876							
Fitness orientation (FO)	r	-.40**	.54**	.21	.21	1					
	Sig. (2-tail)	.005	.000	.144	.145						
	Bonferonni	0.201	0.003#	3.456	3.335						
Health evaluation (HE)	r	-.26	.39**	.10	.18	.45**	1				
	Sig. (2-tail)	.067	.005	.488	.198	.001					
	Bonferonni	2.010	0.236	5.368	3.969	0.057					
Health orientation (HO)	r	-.31*	.53**	.29*	.14	.64**	.47**	1			
	Sig. (2-tail)	.032	.000	.044	.336	.000	.001				
	Bonferonni	1.216	0.005#	1.496	5.376	0.002#	0.033#				
Illness orientation (IO)	r	-.01	.29*	.05	.34*	.31*	.10	.51**	1		
	Sig. (2-tail)	.899	.039	.721	.016	.031	.467	.000			
	Bonferonni	1.798	1.443	5.017	0.656	1.209	5.604	0.010#			
Body area satisfaction (BAS)	r	-.28*	.73**	-.11	.25	.29*	.21	.27	.09	1	
	Sig. (2-tail)	.048	.000	.432	.075	.042	.134	.059	.511		
	Bonferonni	1.584	0.000#	5.616	2.400	1.470	3.484	1.711	5.110		
Overweight preoccupation (OP)	r	.16	-.07	.43**	-.13	.05	-.19	.07	.13	-.39**	1
	Sig. (2-tail)	.270	.619	.002	.351	.728	.198	.613	.357	.007	
	Bonferonni	4.860	4.952	0.094	5.265	4.368	4.158	5.517	4.998	0.294	
Self-classified weight (S-CW)	r	.70**	-.44**	-.04	-.04	-.26	-.21	-.20	.02	-.50**	.28
	Sig. (2-tail)	.000	.002	.773	.768	.074	.136	.173	.866	.000	.053
	Bonferonni	0.001#	0.073	3.092	3.840	2.294	3.400	3.806	2.598	0.010#	1.484

Table 2: Correlations.
(* Correlation is significant at the 0.05 level (2-tailed))
(** Correlation is significant at the 0.01 level (2-tailed))
(# Significant after Sequential Bonferonni corrections)

A positive correlation was found between appearance evaluation and fitness orientation (0.542) as well as between appearance evaluation and health orientation

(0.533). The participants that were generally more positive about their appearance placed a greater importance on fitness and health. This could also indicate that some of the participants exercise for appearance-related reasons, instead of health and fitness.

A negative correlation was found between body area satisfaction and self-classified weight (-0.506). This indicates that the more participants perceive themselves to weight, the less satisfied they would be with those body areas that are associated with being overweight. It could mean that the more unhappy or dissatisfied a person is with her body, the more she is inclined to resort to dieting and eating restraint and perceiving or labeling herself negative. Some of the participants seemed to be unhappy with their appearance and perceived themselves to be larger than they were.

Positive correlations were found between fitness orientation and health evaluation (0.453) and health orientation (0.648). Those athletes who spend more time trying to be physically fit and athletically competent were more health conscious and committed to lead healthy lifestyles. They also perceived their bodies to be in better health than those athletes who spent less time and effort on their fitness.

A positive correlation was found between health orientation and health evaluation (0.472) and health orientation and illness orientation (0.511). The more time and effort athletes seem to invest in leading a healthy lifestyle the more they felt that their bodies were in good health. These athletes also tend to be more alert to personal symptoms of physical illness and are apt to seek medical attention.

	Actual status according to fat percentage		
	Under weight	Normal	Overweight
Total	26	18	5

Table 3: Actual status according to fat percentage.

Two weight categories (underweight: n = 26 and normal or overweight: n = 23) were established according to fat percentage (see Table 3). According to Gore (2000), the recommended minimum and maximum fat percentage for females is between 20 and 25 percent. The groups were therefore categorized according to the following criteria:

- underweight if their fat percentage was less than 20.
- normal if their fat percentage was between 20 and 25.
- overweight if their fat percentage was greater than 25.

Twenty six of the 49 athletes who were tested were considered underweight according to actual status of fat percentage, while 18 athletes were considered normal and five overweight according to their actual status of fat percentage. The normal and overweight categories were grouped together, hence 23 athletes were classified as either normal or overweight according to fat percentage (in one group), while 26 athletes were classified as underweight according to fat percentage (in a second group).

	t	df	Sig. (2-tailed)
Appearance evaluation	1.915	46	0.062
Appearance orientation	1.633	46	0.109
Fitness evaluation	0.529	46	0.599
Fitness orientation	2.949	46	0.005
Health evaluation	1.234	46	0.223
Health orientation	2.386	46	0.021
Illness orientation	1.130	46	0.264
Body area satisfaction	0.941	47	0.352
Overweight preoccupation	0.930	45	0.357
Self-classified weight	−4.321	47	0.005

Table 4: T-tests.

A statistically significant difference (p-value <0.05) between the two groups was found in terms of Fitness orientation (p = 0.005), Health orientation (p = 0.021) and Self-Classified Weight (p-value = 0.005) as measured by the subscales of the MBSRQ (see Table 4). The underweight group had a more positive perception towards Health orientation and Fitness orientation and classified their own weight lower than that of the overweight group. This indicates that the underweight group perceived their health and fitness orientation to be better than those of the normal and overweight group. These results correlate with that of two other studies. Hubby and Cash (1997) illustrated that marathon runners had a greater health and fitness orientation than a normative sample, yet were less preoccupied with body appearance, while Davis and Fox (1993) observed that high-level female exercisers reported greater body satisfaction than lower-level exercisers.

5. CONCLUSION

The correlations between the various subscales of the MBSRQ and the BMI discussed above, give an indication of how the female athletes perceived their bodies, which ultimately defined their perceived body image. This research illustrated that Body Mass Index, Appearance evaluation, and Fitness and Health orientation proved to be the most influential in determining the athletes' perceptions of their body image.

Body Mass Index (BMI) is seen as an indicator of body size. The participants in this sample who felt that they were possibly overweight, also tended to be apathetic towards their physical fitness and health and portrayed a poor body image in comparison with those subjects who did not see themselves as overweight. Previous studies indicated that dissatisfaction with weight-related aspects of one's body (Bercheid & Walster, 1978) and women's feelings about their weight (Cash, 1985) were central to their overall body satisfaction. A perception among some of these athletes that they were overweight, could indicate dissatisfaction with their body size, even though they ran the Comrades Marathon and the mean weight for the sample was 61.28 kilograms (with the maximum weight being 79 kilograms) (see Table 1).

Appearance evaluation portrays the participants' feelings of physical attractiveness or unattractiveness and satisfaction or dissatisfaction with their appearance. The results indicated that those participants who were happy with their physical appearance were also happy with their body size and various areas of their bodies, while those who were unhappy with their appearance were dissatisfied with various body parts and tended to perceive that they were overweight. A study by Cash and Henry (1995) indicated that a large percentage (48%) of women tended to evaluate their appearance negatively. The group of dissatisfied athletes in this sample who seemed to engage in more extensive grooming behavior, was more preoccupied with weight and tended to use dieting and eating restraint as ways to prepare their bodies for participation in the marathon. Many of the participants seemed to participate as a way to improve their appearance. Research (Cash, Novy & Grant, 1994; Davis & Cowles, 1991; McDonald & Thompson, 1992) has shown that such individuals may have a greater chance of having body image distortions, hence they think distortively about their appearances.

Fitness and Health orientation refer to the amount of time spent trying to be physically fit, athletically competent and leading a healthy lifestyle. The participants who perceived themselves to be overweight tended to be more apathetic towards their health and valued fitness less important than those athletes who were underweight. Previous research showed that physically active people had better body images than physically inactive people (Loland, 1998), improved fitness had a beneficial affect on self-esteem and body image (Davis & Fox, 1993) and an active orientation toward fitness was associated with favorable evaluations of appearance and better psychological well-being (Folkins & Sime, 1981). This study showed a less positive attitude to fitness and health from the participants who perceived themselves to be overweight, and therefore experienced a more negative body image than the participants who were underweight. This could mean that they, perhaps, didn't regularly incorporate fitness and health activities in their lives, although they participated in the Comrades Marathon.

Although the results indicated that there was a correlation between the perceived and observed body image of female Comrades Marathon athletes, there was also a statistically significant difference between the athletes who were underweight and those who were normal or overweight. It seems that those athletes who perceived themselves to be overweight tended to struggle more with their body image, while those who were underweight were more satisfied with their body image as marathon runners. Gill (1999) stressed that females in certain sports may have exaggerated body image concerns related to appearance and performance and encourages sport psychology consultants to take a more active feminist approach to try and change the system that leads athletes to pursue an unhealthy body image.

Body image is closely tied to the notion that a person's feelings of self-worth are related to how she perceives herself. A positive body image contributes to the development of global self-esteem. Athletes who enjoy a high level of positive body image are more likely to enter into competition and feel good about themselves and exercising. This research illustrated that BMI, appearance evaluation as well as fitness and health orientation are influential in determining these athletes' perceptions of their body image. Female athletes who perceives themselves to be either underweight or overweight and therefore struggling with their body weight, should therefore receive guidance from professionals, like sport psychologists to assist them in this regard.

REFERENCES

Bercheid, E. & Walster, E.H. (1978). Interpersonal attraction (2nd ed.).Reading: Addison-Wesley.

Brenner, J.B. & Cunningham, J.G. (1992). Gender differences in eating attitudes, body concept, and self-esteem among models. Sex roles, 27, 413-437.

Carter, J.E.L. (1982). Body composition of olympic athletes. In J.E.L. Carter (Ed.), Physical stature of Olympic athletes, Part 1, Montreal Olympic Games Anthropological Project. Basel: Karger.

Cash, T.F. (1985). Physical appearance and mental health. In A. Kligman & J.A. Graham (Eds.), The Psychology of Cosmetic Treatments. New York: Preager Scientific.

Cash, T.F. (1994a). Body image and weight changes in a multisided comprehensive very-long-calorie diet program. Behavior Therapy, 25, 239-254.

Cash, T. (1994b). Multidimensional Body-Self Relations Questionnaire: MBSRQ User's Manual. Norfolk: Old Dominion University.

Cash, T.F. (1996). The treatment of body image disturbances. In J.K. Thompson (Ed.), Body image, eating disorders, and obesity: An integrative guide for assessment and treatment (pp. 83-107). Washington D.C.: American Psychological Association.

Cash, T.F. & Henry, P.E. (1995). Women's body images: the results of a national survey in the USA, Sex Roles, 33, 19-28.

Cash, T.F. & Hicks, K.L. (1990). Being fat versus thinking fat: relationships with body image, eating disorders, and well-being. Cognitive Therapy and Research,14, 327-341.

Cash, T.F., Novy, P.L., & Grant, J.R. (1994). Why do women exercise?: Factor analysis and further validation of the reasons for exercise inventory. Perceptual and Motor Skills, 78, 539-644.

Cash, T.F. & Pruzinski, T. (1990). Body images: development, deviance, and change. New York: Guilford Press.

Cash, T.F., Winstead, B. & Janda, L. (1986). The great American shape up. Psychology Today, 2, 30-37.

Clark, N., Nelson, M. & Evans, W. (1988). Nutrition education for elite female runners. The Physician and Sportsmedicine, 16, 124-136

Davis, C. & Cowles, M. (1991). Body image and exercise: a study of relationships and comparisons between physically active men and women. Sex Roles, 25, 33-44.

Davis, C. & Fox, J. (1993). Excessive exercise and weight preoccupation in women. Addictive Behaviors, 18, 201-211.

De La Rey, C. & Paruk, Z. (1993). A psycho-demographic survey of women runners. South African Journal for Research in Sport, Physical Education and Recreation, 16(1), 25-33.

Drewnowski, A & Yee, D.K. (1987). Men and body image: are males satisfied with their body weight? Psychosomatic Medicine, 49, 626-634.

Folkins, C.H. & Sime, W.E. (1981). Physical fitness training and mental health. American Psychologist, 36, 373-389.

Garner, D.M. & Rosen, L. (1991). Eating disorders among athletes: research and recommendations. Journal of Applied Sport Science Research, 5, 100-107.

Gill, D.L. (1999). Gender issues: making a difference in the real world of sport psychology. In G.G. Brannigan (Ed.), The sport scientists: research adventures. New York: Longman.

Gore, C.J. (2000). Physiological tests for elite athletes: Australian Sport Commission. Champaign: Human Kinetics.

Harris, M.B. & Greco, D. (1990). Weight control and weight concern in competitive female gymnasts. Journal of Sport and Exercise Psychology, 12, 427-433.

Hollilman, S. (1991). Eating disorders and athletes. New York: Kendall Publishing.

Hubby, D.C. & Cash, T.F. (1997). Body image attitudes among male marathon runners: a controlled comparative study. International Journal of Sport Psychology, 28, 227-236.

Humphries, L. & Gruber, J.J. (1986). Nutrition behaviors of university majorettes. The Physician and Sports Medicine, 14, 91-98.

Loland, N.W. (1998). Body image and physical activity: a survey among Norwegian men and women. International Journal of Sport Psychology, 29, 339-365.

McDonald, K. & Thompson, J.K. (1992). Eating disturbance, body image dissatisfaction, and reasons for exercising: Gender differences and correlational findings. International Journal of Eating Disorders, 11, 289-292.

Rosen, L.W., McKeag, D.B., Hough, D.O., & Curley, V. (1986). Pathogenic weight-control behavior in female athletes. The Physician and Sport Medicine, 14, 79-86.

Soap, R.A. & Murphy, S.M. (1995). Eating disorders and weight management in athletes. In S.M. Murphy (Ed.), Sport Psychology Interventions. Champaign: Human Kinetics.

Weight, L.M. & Noakes, T.D. (1987). Is running an analog or anorexia? A survey of incidence of eating disorders in female distance runners. Medicine and Science in Sport and Exercise, 19, 213-217.

Wells, C.L. (1991). Women in sport and performance. Champaign, IL.: Human Kinetics.

Zulu, N. (1999). We can beat the world if they give us a break. Pace, Jun., 1999, 62-63.

Index

Adipose tissue (mass)   54–7
Adiposity   33, 35, 55–7
Androgyny index   57
Anthropometry   1, 40, 53, 54, 57, 59,
  61, 71, 75, 95, 96, 98, 105, 128, 132,
  133, 138–41, 146, 148

Bioelectrical impedance analysis
  (BIA)   65–9, 72, 73, 80–4, 86–9,
  95–9, 104, 105, 109, 112
Body builders   15, 17–26
Body composition   23, 29, 39, 53–5,
  61, 95, 96, 100, 104, 105, 110,
  112, 124, 126, 129, 131–6,
  139, 141, 148
Body density   41, 133
Body image/perception   26, 131,
  132, 138, 139, 141, 145–7,
  151, 152
Body mass   16, 33, 35, 36, 40, 47, 53,
  71, 73, 99, 112, 128, 140
Body mass index (BMI)   58, 75, 96,
  131–4, 136–40, 147, 148, 151
Bone mass   54
Bone mineral density   124–6, 129
Brachial index   32, 57
Breadth(s)   19, 20, 24, 33, 44, 47–9,
  57, 133

Cadaver dissection   54
Capacitance   67, 96
Computed tomography (CT)   61
Crural index   57

Densitometry   54, 55
Density   54
Dual X-ray absorptiometry (DXA or
  DEXA)   53, 54, 58–61,
  123–30, 140

Extra cellular water (ECW)   65, 67, 68,
  81, 82, 85, 87, 89

Fat free mass   54–6, 97, 99, 100, 101, 104,
  112, 125, 132–6
Fat free soft-tissue mass   125, 126
Fat mass   16, 25, 45, 49, 54, 55, 57, 97,
  123, 126
Fat patterning   57
Fat, body fat   16, 25, 55, 60, 61, 96, 109,
  111, 112, 115, 117, 123, 124–9, 131, 132,
  139, 140

Girth(s)   16, 19, 20–2, 24, 33, 43, 133
'Gold standard'   54
Gymnastics, gymnast(s)   109–12,
  114–18

Heterogeneous   34
Homogeneity   29, 30, 33, 99
Hormonal levels   109, 111, 115
Hydrodensitometry, hydrostatic
  weighing   123, 129
Hypohydration   68, 70, 73, 75, 81, 83–6,
  88, 89

Impedance   66–70, 72–83, 85, 86, 88,
  89, 96, 104
Impedivity   66
Infrared thermometry   74
Interference   3
Intra cellular water (ICW)   65, 81,
  82, 85, 87
Isotopic dilution   65, 84

Landmarker   7, 11
Lean tissue mass, lean body mass   16, 116,
  123, 128
Length(s)   19, 20, 23, 33, 44, 48, 57, 71

Magnetic resonance imaging
  (MRI)   54, 61
Maturation   109–12, 114–18
Menarché   109–11, 113, 115, 117
Morphological optimisation   54

Morphological prototype   54
Morphology   15, 29, 30, 34, 36, 39, 40,
   45, 49, 53, 55, 56, 59, 62
Muscle dysmorphia   131
Muscle mass   25, 41, 45–7, 49, 54, 56, 57,
   59, 61, 124, 131, 140

Obesity   134
Oestrogen, Estradiol   109, 112–13,
   115, 117
Optimization, 31
Overlap zone   31
Overweight   137–8, 148, 150–2

Percentage body fat, %fat   41, 45,
   55, 82, 95, 100, 102–4, 111–12,
   115, 124–5, 129, 133, 135,
   137–40, 147
Phantom Z scores   16, 20, 21, 23–4
Phenotype   53, 56, 57
Physique   53, 56
Progesterone   109, 112, 113, 115, 117
Proportion, proportionality   32, 53
Puberty   110

Reactance   67, 68
Residual mass   54
Resistance   68, 83, 96
Resistivity   66, 81–3, 85, 87

Secular trend   31
Sexual dimorphism   58
Shape   29, 53
Size   29, 53
Skeletal mass   41, 45–6, 49, 54
Skelic index   57
Skinfold(s)   16–7, 20, 24–5, 35–6,
   41, 49, 57–8, 60, 98, 105, 123–30,
   133–7, 140–1, 147
Somatotype   16–18, 26, 30, 41,
   46, 49, 104
Stature   35–6
Strongmen   15–26, 61
Subcutaneous adiposity, subcutaneous
   fat   15
Surface area   7
Symmetry   53, 59

Talent identification   29, 30, 36
Testosterone   110, 112–13, 115–17
Three dimensional (3D) scanner   1–3, 6,
   8, 9, 12, 61
Total body water (TBW)   65, 67–9, 74, 77,
   80–5, 87–9, 104

Ultrasound   61
Underweight   138, 148, 150–2

Whole body scanning   5